Written, edited and designed by the editorial staff of ORTHO BOOKS.

Food Editor:
Annette C. Fabri

Illustrations:
Ron Hildebrand

Photography:
William Aplin
Clyde Childress
Michael Landis

All About Tomatoes

Contents

The tomato and its habits

After its long hard trip from obscurity, criss-crossing continents, the tomato has become the No. 1 plant in our gardens today. Gardeners have glorified it, plant breeders have improved it. Yet it is unpredictable in growth habit and the perfect tomato is a matter of personal preference.

In our deep involvement with the tomato there have been disappointments, but there have been great tomato-filled moments in which all frustrations are forgotten.

During the birth of the tomato there are moments of awe. Even before its true leaves are formed, the seedlings seem to have unusual determination, and willingness to grow. They promise a great future.

And then there's the day when the seedling, wider than tall, conditioned by daily trips outdoors, is ready for its permanent place in the garden. You listen to the weather report and wonder if the garden weather is ready for your perfect transplant.

Then the day of your first ripe tomato. No tomato can equal it in flavor. This is probably the greatest tomato-filled moment of the year. This is the moment you've been waiting for. More than a reward for months of labor, it signals the beginning of the annual festival of home-grown tomatoes ripening on the vine.

In the days that follow, you check the vines, heavy with green fruits, and wish them good sailing through the season ahead. Are the new flowers setting fruit? Will each vine keep its promise?

Then the daily treasure hunt. The daily visit with the vines, blessing those that give you ripe fruit, encouraging those with green fruit, always with questioning eyes for symptoms of disease.

◁

The life of the variety 'Springset,' from seed, to seedling, to maturity, to ripe on the vine.

As the harvest season picks up, the kitchen becomes a most important part of the garden. Suddenly menus begin to change, revolving around the juicy, fresh tomatoes which abound in the garden. Salads become zestier, sauces richer, casseroles take on a "spruced up" look and taste. There's never a question about a fresh vegetable for dinner. Finger foods can be picked right from the vine and cooking becomes a delightful challenge to the cook. Things begin to happen in the kitchen that couldn't happen before.

As you are enjoying these days of the home-grown tomato you may wonder why you can't buy this fruit. The answer is simple:

Something good happens to the tomato that is ripened *on the vine.*

The same good things don't happen to the tomato that is ripened *off the vine.*

That is the one big difference between store-bought and home-grown tomatoes.

Let's play fair with the tomato

The supermarket fresh tomato has received much damaging publicity. To the zealous critics, the growers of tomatoes have conspired to palm off inedible cannonballs.

The truth is this:

1. The quality of the supermarket fresh tomato is improving.
2. Growers of fresh market tomatoes are financing studies on every phase of post-harvest handling of the tomato.

The quality of the supermarket tomato is the concern of the tomato breeder, growers, and all of us. The season for fresh tomatoes in the store is every month in the year. The season

"Just look at my tomatoes."

The daily treasure hunt.

In six-inch pots in a sunny window.

The 'Sweet 100' is sweet.

As beautiful as any vine.

Rough, but full of flavor.

we can enjoy the home-grown tomato is a short three months—more or less.

The varieties of tomatoes in the market are the same varieties that are praised in the home gardens in the regions where they are grown. The soil they are grown in must contain all of the nutrients necessary for fruit production.

The measurement of quality changes when you buy fresh tomatoes. Home gardeners have a far different standard when buying tomatoes than when growing tomatoes. When we grow tomatoes, abnormal shapes are part of the fun and a crack or two is quickly cut away. When we buy tomatoes we choose those that are uniform in size, smooth, and free of all blemishes.

The tomato is a perishable fruit. Once the ripening process starts in, it can't be completely stopped. It ripens rapidly at temperatures of 72° F., stores well at 60° F., can be held at 55° F.

The tomato ripens from the inside out. The tomato that is called vine ripe is picked from the vine when pink color first shows at the blossom end. (See page 95.)

The day will come when we can buy a tomato that is ripened on the vine. Its skin will be tough enough to withstand the rigors of shipping. But it will be easy to peel. It will be firm but

The tomato ripens from the inside out.

Ready for green tomato pie.

Ripening on the vine, in the garage.

The juice of the vine.

not rock firm, with the proper sugar-acid ratio that most palates call good flavor.

Green is the color

In the course of our affair with the tomato and our infatuation with the red, ripe ones, the pink, yellow, white, and green were not forgotten. All tomatoes are wonderful.

It was at the end of the season when we were faced with a glut of green tomatoes that we discovered the all-important place of the green tomato in the kitchen. Rather than hanging the vine with green fruits upside-down in the garage to ripen or rot, we channeled them right into the kitchen.

We expected relishes and piccalilli, but the surprises were the pan-fried slices of green tomato, green tomato jam and marmalade, and best of all the green tomato pie.

Avoid trouble

In the following pages we put the tomato through all its paces. Remember this as you read: if you have had trouble growing tomatoes, favor the disease-resistant varieties.

There are, of course, areas that are free from verticillium wilt, fusarium wilt, and nematodes, but again *if you had trouble* growing tomatoes, favor the disease-resistant varieties.

It's had a long hard trip

This is a brief, informal account of the origin of the cultivated tomato, as it seems to stand at the present time. It was prepared for us by Dr. Charles M. Rick, a professor in the Department of Vegetable Crops, University of California, Davis. Dr. Rick works primarily on basic tomato genetics and is one of the leading authorities in this field.

The origin of the cultivated tomato is shrouded in mystery. Considering the perishability of all parts of the tomato plant, it isn't surprising that archeology hasn't been much help in telling the story. However, we can piece together what we know about the tomato historically—where it grew, who grew it, how it was used. And we also have helpful genetic evidence. Definite origin of the tomato may defy us, but we can work out a fairly reasonable conjecture.

What, then, are the solid facts? First, the genus *Lycopersicon*—the botanical group to which the tomato belongs—is native to western South America, and only *Lycopersicon esculentum* var. *cerasiforme,* the wild cherry form of the cultivated species, has spread throughout Latin America and the Old World Tropics. Second, the tomato was not known in Europe until *after* the discovery and conquest of America, descriptions and drawings first appearing in the European herbals of the middle and late 16th century. Third, these writings clearly reveal that man had been trying to improve the tomato with regards to size and diversity of shape and color. These improvements over the wild ancestors were almost certainly achieved by primitive man in America. Mexico appears to have been the site of domestication and the source of the earliest introductions, and the wild cherry tomato was probably the immediate ancestor.

According to this hypothesis, wild, tiny, red-fruited tomatoes of Peruvian origin migrated northward through Ecuador, Colombia, Panama, and Central America to Mexico, where the migration seems to have stopped. We assume this because we have no evidence that North American Indians

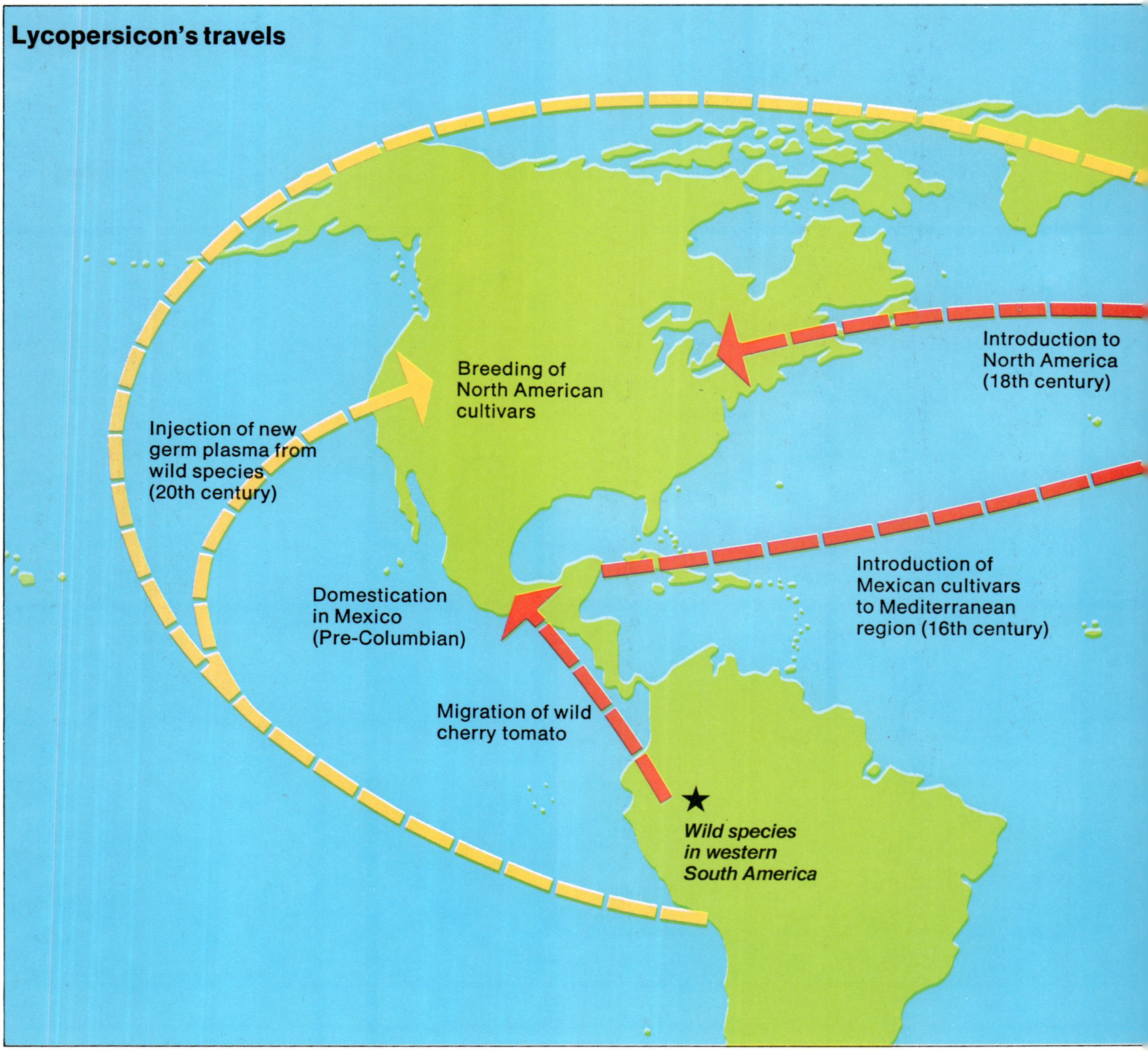

were acquainted with the tomato. In Mexico, man's continued selection for larger and more diverse fruits led to domestication. Recent research on hereditary characters of wild cherry tomatoes from this route reveals that some evolutionary process occurred *during* the move north from Peru. We know this because genetic tests show that old European and many modern cultivars have the same genes as these migrant cherry types, but differ considerably from those of the wild and cultivated tomatoes from Peru and Eucador.

After the trip from Peru to Mexico, then where? The map illustrates the wanderings of the tomato. Introduction to Europe, possibly first in Spain, was followed by slow migration throughout the Mediterranean region, and later to northern Europe. The tomato was first grown as a curiosity, or possibly for the ornamental value of its fruit. Its culinary use was delayed by fears of poisonous qualities. The tomato belongs to the Nightshade family, which includes some plants of well-known toxicity. By association with them, fear spread to the tomato and persisted even into this century—long after its food value was discovered by the Italians and other Europeans.

As with the potato (also of American origin), the tomato was reintroduced to the New World from Europe. In the United States the tomato was of minor importance until the middle 1800's, although Thomas Jefferson was growing it in the late 18th century. The first known attempts to breed improved types date from 1860. Until the turn of the century, most of the improvement resulted from selection within existing stocks, mainly of French origin. As the tomato gained importance as a food plant, breeding efforts accelerated, with extensive hybridization and selection. Until modern times, breeders limited their efforts to working with European stocks as parent plants. These proved to be restricted sources when the need arose for resistance to various pests and for other characteristics not present in European stocks. Breeders then exploited genes in primitive cultivars and wild species from the native region. The resistance to fusarium and verticillium wilt and to nematodes, now incorporated in many cultivars, are products of such efforts.

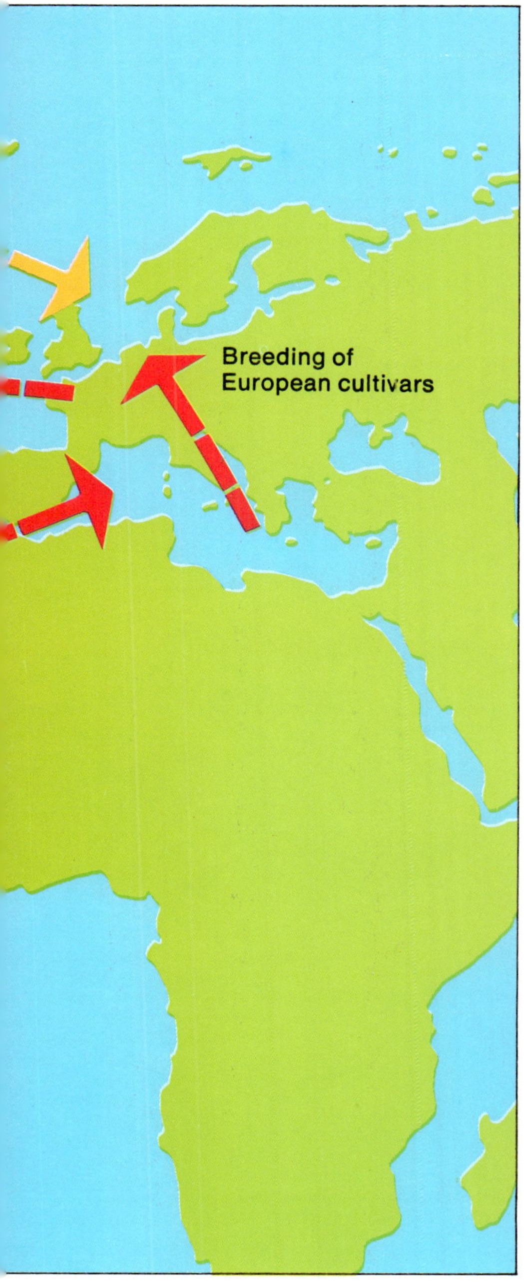

The plant, flowers and fruit of Lycopersicon esculentum *var.* cerasiforme, *the theoretical ancestor of all modern tomato varieties.*

Anatomy of the tomato

Determinate tomato

Terminal buds do not set fruit; stop stem growth. The plant is self-topping and seldom needs staking.

All the blossoms and fruit develop on a plant at about the same time. Harvest time is short, only a week to ten days for some varieties.

Some anatomical details

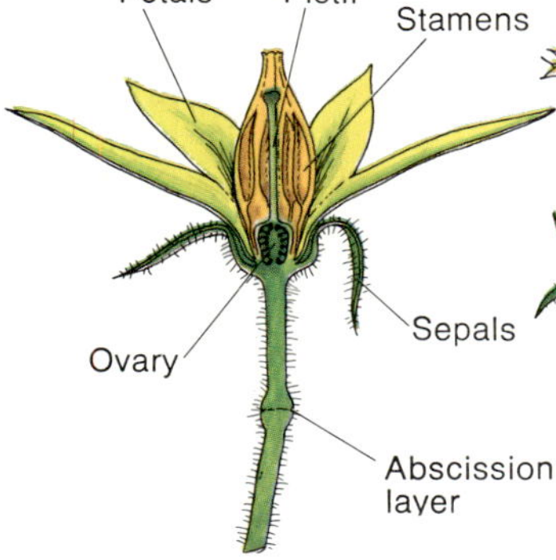

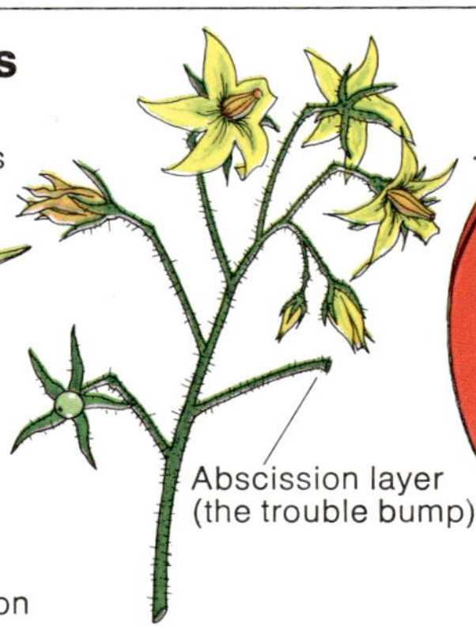

Tomato flowers are complete with both male and female organs and are mostly self-fertilizing. When fruit fails to set and blossoms drop the abscission layer is where it happens (see page 10).

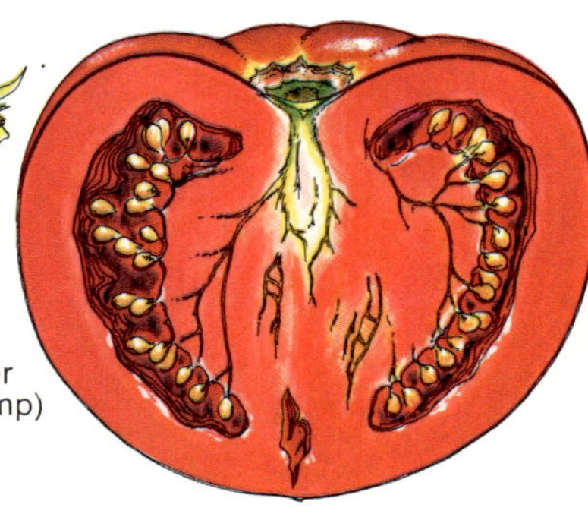

The fruit has two or more chambers called "locules." Large-fruited varieties have 5 to 10. A gelatin-like substance surrounds the seeds.

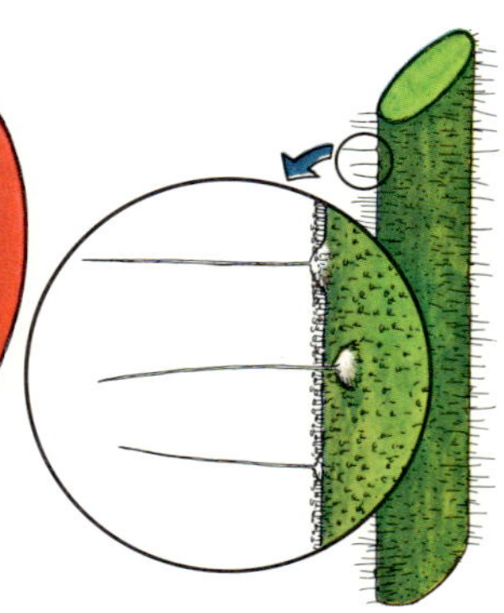

Glandular hairs are found on stems and leaves. When bent they give off the oil that give tomatoes their characteristic odor.

Indeterminate tomato

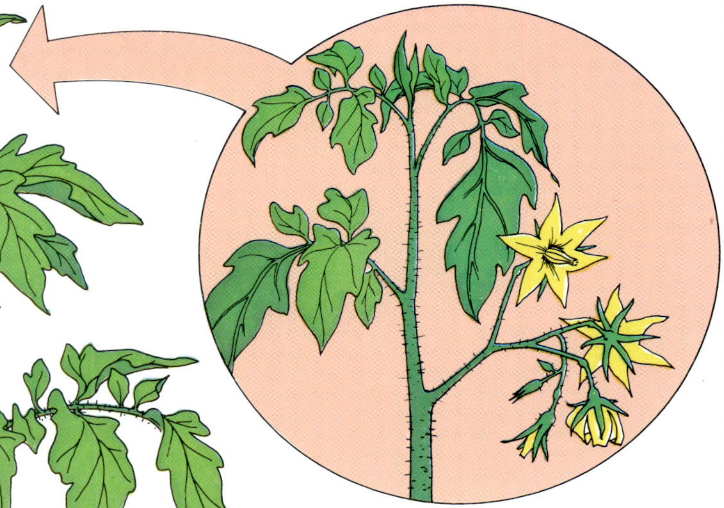

The terminal bud of an indeterminate tomato plant does not set fruit, it always produces leaves and more stem from the growing tip. The vine can grow indefinitely if not killed by frost in the fall.

Also the blossoms and fruit develop progressively as the vine grows so tomatoes in all stages of development may be on the vine at one time and the harvest may last for several months.

NOTE: This drawing and those on the opposite page have been simplified and modified for clarity. Reasons and recommendations for pruning to one, two, or more stems can be found on page 28.

Standard pollination

POLLINATION: Pollen grains are released by the anther and some fall or float in the air to the stigma, usually of the same flower, where they adhere to the sticky surface.

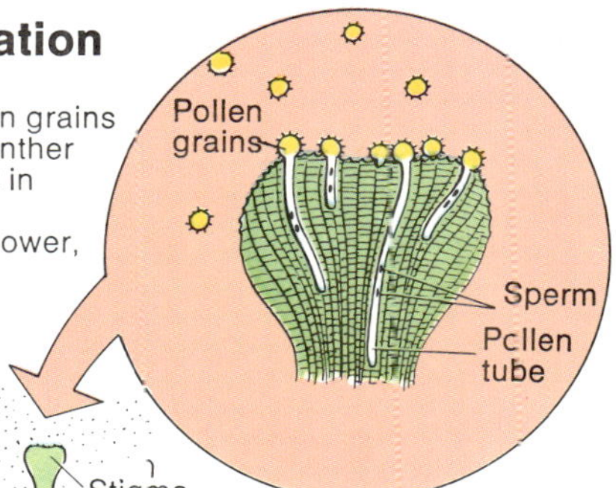

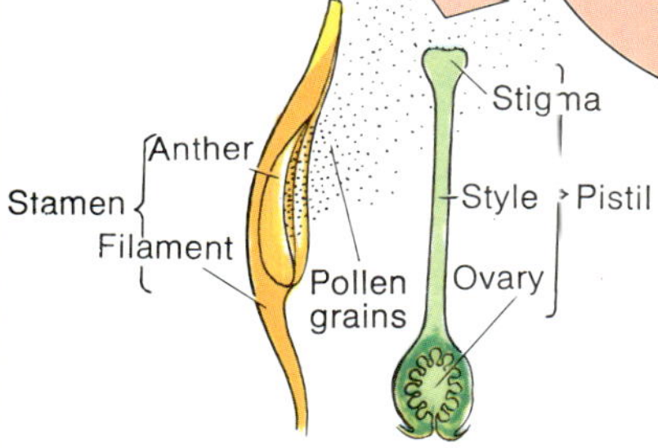

FERTILIZATION: When pollen grains germinate and send tubes with the pollen tube nucleus (sperm) down the style where they unite with the ovules, in the ovary, the fruit is set.

Hybrid pollination

1 Cross pollination is accomplished by removing the anthers of parent X before pollen is released . . .

2. The Y-parent's anther tube is opened up when pollen is present. The pollen is picked up with a soft brush and . . .

3. Deposited on the stigma of parent X.

The seeds from the resulting tomato have characteristics of both X and Y.

Flowers fail to set fruit when they are not fertilized with their own pollen. An abcission layer forms at the node and the blossom will drop off.

Pollinated (fertile) flowers and the fruits that follow remain well secured to plant.

Quirks and myths

The very good gardener who grows prize-winning carrots, beets, chard, beans, and even cauliflower may have trouble growing tomatoes. The tomato is not a stolid, solid citizen. When all growth factors are right, it produces magnificently. The quirks in the behavior of the tomato plant have led to speculation about every stage in its development.

We talked about the quirks in the growth of the tomato plant with Dr. Philip Minges of Cornell. We threw at him the questions most commonly asked about tomatoes. His answers are in quotes. Answers without quotes are based on our research and experience.

Why blossom drop?

Many of the main-season tomato varieties will set fruit only within a rather narrow range of night temperature.

Temperatures above 55°F for at least part of the night are required for the first fruit set. Night temperatures above 75°F in the summer months inhibit fruit set—cause blossom drop.

Most of the early-season varieties will set fruit at lower temperatures. Plant breeders have developed varieties that will continue to set fruit at night temperatures above 75°F.

There are good years and bad years for tomato growing in every section of the country. Spring weather may be erratic, abnormally cold and wet. A heat wave may come along in July to knock the blossoms off.

To many a gardener the most critical hours in the life of a tomato plant are the hours it is setting fruit. The blossoms are out. The big question: "Will they drop or set fruit?" There's a wait of about 50 hours to find out. It takes that long or longer for the pollen to germinate and the tube to grow down the pistil to the ovary. At night temperatures below 55°F the germination and tube growth is so slow the blossoms drop off before they are fertilized.

Why all vine and no fruit?

Some time ago we wrote that, given too much nitrogen and water in the early stages, a tomato plant might go right on producing foliage at the expense of fruit.

We were wrong.

No one can question the fact that a heavy application of nitrogen and water *will* produce vigorous vine growth—but that isn't the whole story, as a scientist pointed out to us.

Actually, here's what happens. Once the flowers set, the plant takes care of itself. If you feed it lots of nitrogen, and flowers don't set, then you get a lot of vine growth. But nitrogen is not responsible for a tomato plant's failing to change gears from the vegetative to the fruiting stage.

The plant is all vine and no fruit simply because no fruit set. When the blossoms drop, a plant continues in the vegetative state.

Here's an example of "all vine and no fruit" in our own garden.

We have two plants of 'Fantastic,' two of 'Delicious,' and two of 'Yellow Plum.' We are picking ripe fruit from the first blossom cluster of 'Fantastic. But all the early blossoms of 'Delicious' dropped, and fruit is only now setting on flowers—at five feet above ground. Then there is 'Yellow Plum' at seven feet, and producing from the *ground up.*

What's the explanation? Night temperature—not too much nitrogen. Fertilizer treatments were the same on all three varieties. Evidently, the 'Delicious' needs a higher night temperature than the 'Fantastic' for first blossom set. For now 'Delicious' is a beautiful vine with two green fruit, maybe more tomorrow.

Can I improve fruit set?

Pollen is shed most abundantly on bright sunny days between 10 A.M. and 4 P.M. It occurs within individual flowers. Tapping or jarring the entire plant is only effective when the flower clusters are near the top of the plant. To increase pollination from top to bottom, give individual attention to all flower clusters with a daily vibration, using an electric vibrator or battery-operated toothbrush. Best time to shake and vibrate is in midday when it's warm and the humidity is low.

Does sprinkling knock off flowers of tomatoes and other vegetables?

"No, not unless the flowers were about to fall off anyway. When flowers fail to set, in a few days an abcission layer forms at a node, as with leaves in autumn, and in a week or so they will drop off. Fertilized flowers and fruits remain well secured to the plants so that it is almost impossible to pull or knock off (though there is an occasional exception)."

Will sprinkling in hot weather cause leaf burn?

"No. If anything, the plants will be cooled and the transpiration rate will drop." It is a myth that drops of water on the leaves act as a prism to magnify the heat.

Will crowding tomato plants reduce fruit size?

"Usually not enough to be perceptible. Crowding of this group of vegetables usually reduces the yield per plant rather than the size of fruit. Adequate soil moisture during the fruit-sizing period is a major factor for obtaining normal size of fruit.

One way to pick ripe tomatoes is to remove them at the nodes.

One method to increase pollination is to give each flower cluster a daily vibration. An electric toothbrush works well.

Will pruning tomatoes help or hinder good yield?

"Tomato yields per plant may be lowered by pruning. Removing the leaves or shoots does not conserve 'food' for the crop, it tends to reduce the total food supply. With certain methods of staking, some pruning is necessary, but when it is possible, use training methods that require little pruning."

Is direct sunlight necessary for ripening tomatoes?

"No. Tomatoes will turn color and ripen in shade or in the dark at proper temperatures. In fact, red or pink tomatoes will turn yellow on the exposed side if they receive too much sun, so some foliage cover is beneficial. Dense foliage late in the summer or early fall may retard ripening because of lowering the fruit temperature."

Sometimes tomato fruit have clean holes in the side, associated with brown "zipper" streaks, or bands. Sometimes the fruit is quite deformed. What is the cause?

"This disorder might be called 'blossom-tear.' It is a growth defect that begins shortly after the setting of a fruit. Blossom-tear seems to occur most frequently when wet, cool weather prevails during the flowering period, causing the corolla (yellow part of the flower), to stick to the small ovary. As the ovary expands, the corolla eventually breaks loose, tearing away a small amount of wall tissue. The hole enlarges as the fruit grows. Blossom-tear basically harms only the appearance of the tomato. No control has been discovered."

A zipper streak on a beefsteak-type tomato.

What is the cause of blotchiness in tomatoes and how can it be controlled?

"The cause is not clearly understood and there is no consistently satisfactory control. Blotchiness is a physiological disorder characterized by lack of red pigment and hard tissue. (Sometimes dark or black strands appear in the white, or greenish, abnormal tissue.) It may be visible externally or apparent only after cutting the fruit. You can use fruits with a little, or moderate amount of blotchy tissue, but badly affected fruits are culls. Though no varieties are immune, some tend to be less affected than others, for example, 'Heinz 1350,' is a less susceptible variety."

How about quality?

The quality of a tomato of any variety is not an inherent characteristic. Quality of the individual fruits depends on a total of all the factors which contribute to the plant's growth: sunlight, soil, moisture supply, nutrient supply, methods of cultivation and exposure—all play a part in determining the ultimate quality of the fruit. Tests made by Cornell University on the variety 'New Yorker' show that its quality will vary slightly from garden to garden, within a mile or two.

Should I worry about leaf rolling on tomatoes?

"In most cases there are no diseases or other disorders involved, and yields are not affected. On many varieties, after a good load of fruit has set, the lower leaves begin to roll up, or cup. We've found that the greater the ratio of fruit to leaf area, the more the leaves will roll. Varieties with a lower ratio of fruit to leaf area, have less leaf roll. Poor-yielding varieties have less roll. Pruning off leaves and shoots will intensify rolling —so will a moisture shortage. A few varieties have a genetic tendency for 'rolling characteristic.' These will often exhibit leaf rolling even before any fruit set."

How come potatoes sometimes produce tomatoes?

"They don't. Some years, potato flowers set to produce potato seed balls, which resemble tomatoes."

Intergeneric grafting is possible within the *Solanaceae* family. We have seen a very strange-looking plant on which eggplant, pepper, tomato and petunia have been grafted.

Should I save seed?

If you want to perpetuate a *standard* variety of tomatoes that has performed well in your garden, save seed from some of its fruits. However, if you're planning on saving seed of any of the *hybrids,* you should know this: The first season you plant the seed of 'Better Boy,' or any other hybrid, you will get a duplication of 'Better Boy,' or whatever hybrid you buy. But, hybrid seed is genetically stable for only one season. After that, seed is unstable. You may get six different varieties from six seeds.

In the list of varieties starting on page 34 we note whether the variety is standard (S), or a hybrid (Hy) variety.

Your favorite tomato?

Is it the small 'Red Cherry' size, just right for salads? Or is it one of the giant beefsteak-type varieties, in search of a hamburger bun?

Or can it be of any size as long as it is ripened on the vine? Perhaps your favorities are many tomatoes.

Sunlight
Sunlight is used by the leaves to convert raw materials into useable plant food.
Oxygen & Water
Carbon dioxide is absorbed into leaves and they give off oxygen and water.
Carbon dioxide
The stems are two-way streets. Raw materials move up into the leaves and the food is manufactured in the leaves and circulated throughout the plant.
Oxygen
Carbon dioxide
Roots require oxygen for growth and they give off carbon dioxide.
Soil nutrients
Nutrients dissolved in water are absorbed by the roots and move up into the leaves.
Water

Grow it

So you want to grow a bumper crop of perfect tomatoes. First of all, it's well to remember that you are only an assistant. The plant does the growing. Your function is to supply the soil, moisture and nutrients it needs. Here's how.

You want perfect tomatoes. The tomato plant is ready and willing to deliver if you give the plant what it needs. It needs what every plant needs—and *more:*

1. A continuous supply of moisture.
2. A continuous supply of nutrients.
3. Air in the soil for healthy root growth.
4. About 8 hours of daily sunlight.
5. Night temperatures (at least during part of the night) that permit flowering and setting of fruit.
6. Protection from insects, diseases, and dogs that dig holes to nap in.

You don't have to be a horticulturist to grow tomatoes or to put the basic rules of plant growth together.

It seems to us that what trips up many a grower of tomatoes is an underconcern with one or another of the needs of the plant. Actually, these six fundamentals are one fundamenta ; *all* must act in unison.

You can go into the garden, look at a tomato plant, and decide that it needs more fertilizer. Fine—but you should bear in mind that fertilizing, the nature of the soil in which the tomato vine is planted, and your watering habits are all interdependent.

If you plant a six-pack of the tomato variety recommended by your garden store, and all plants grow and produce well, you can count yourself lucky. But if your experience hasn't been all that successful, the following pages of advice in words and illustrations may be helpful.

Give it what it needs

A continuous and uniform supply of water: neither too much, nor too little. Too much water can drown plant roots, especially if soil is heavy. Too little water can stop tomato production.

A continuous supply of nutrients: The daily requirements of plants for nutrients is small, but that amount should be available when the plant needs it. Always follow package or bottle directions for best results.

Air in the soil after drainage: The most important factor in the growth of plants. Roots breathe the way you do. Deprived of air, roots, or portions of roots die of suffocation. Make sure soil is loose, friable, permitting good drainage.

Sunlight: To a plant, light is life. Sunlight is used by the leaves to convert raw materials into useable plant food, and provides the energy required for photosynthesis. The tomato should have 8 hours of continuous sunlight, but can get along with less. See pages 54-57.

Temperature: Most tomatoes need night temperatures between 55° and 75° to set fruit. But there are varieties that will set fruit at lower and higher temperatures. See pages 34-53.

Protection of leaves and roots. Leaves and roots need protection from temperature extremes, strong winds, birds and beasts, weed competition, and pests and diseases. But if you've satisfied the first five requirements, the plant is in a good position to protect itself from any damager, and will require much less protection from the gardener.

Starting from scratch

The trick in sowing seeds indoors is timing. You want an ideal-sized transplant to set out in the garden when the weather is just right. Figure on five to seven days for seed germination, eight to ten weeks to grow to transplant size. That means sowing the seed 11 to 12 weeks prior to the optimum planting time in your area. Eight to twelve plants should be enough for both fresh tomatoes and for canning. Since seed is relatively inexpensive, start with more peat pots than are absolutely needed. The increase in cost is minute, but the practice may make the difference between failure and success. You may not get 100% germination—and some seedlings may not make the grade to transplant size.

Direct seeding may be into 3- or 4-inch clay or peat pots or some of the newer plant-growing containers. Fill the pots to about ½-inch from the top with a sterilized potting medium. Plant one to three seeds ⅓- to ½-inch deep in the center of each pot. Plant at a uniform depth so seeds will germinate evenly and result in uniform plants for transplanting into the beds. After germination, thin seedling to one per pot.

Whether in trays, cubes, or what-have-you, the material should be thoroughly moist before seeds are sown. Give them the warmest spot you can find. For fast germination, seeds of tomato need a soil temperature of 75° to 85° F.

Cover blocks or peat pots or trays with paper or slip them into a plastic bag to prevent drying out. There should be no need for watering until the seeds have sprouted.

Once the seedlings emerge, keep them in full sunlight—12 hours a day if possible. The temperatures for seedling growth should be between 70° and 75° F. during the day and 60° to 65° F. during the night.

Data compiled by the Department of Vegetable Crops, University of California at Davis, show the following rates of germination at different temperatures:

Days to appearance of seedlings at various soil temperatures from seed planted at a half inch depth.

	Soil Temperatures in Degrees Fahrenheit					
Crop	50	58	68	77	86	95
Beans, Snap	0	16	11	8	6	6
Beans, Lima	0	30	18	6	6	X
Beet	17	10	6	5	4	4
Cabbage	15	9	6	4	3	0
Carrot	17	10	7	6	6	8
Corn	22	12	7	4	4	3
Cucumber	X	13	6	4	3	3
Lettuce	7	4	3	2	2	X
Muskmelon	0	0	8	4	3	0
Pepper	X	25	12	8	8	9
Radish	11	6	4	3	3	0
Tomato	43	14	8	6	6	9

X = little or no germination 0 = not tested

Seeds planted deeper than one-half inch take more days to appear than shown in this chart. Soil temperatures are slightly cooler as you go deeper and the seedlings have a greater distance to grow before appearing at the surface. Many seeds, when planted too early in cold soil, will rot before the temperature is right for germination.

To be sure the soil in your garden is warm enough for planting test it with a soil thermometer. By taking soil temperatures all over your garden you may find some areas warming earlier than others. These are the areas you should plant in first.

Containers for seed starting

SOIL MIXES. Packaged potting soil is a sterilized and scientifically formulated medium that provides proper water retention and drainage for your new plants. Having strong, healthy seedlings with no insects and no damping-off or other diseases is well worth the extra cost.

FLATS. New plastic flats and the old-fashioned wood ones are still used extensively for seed starting (see "Two-step Method" on page 17).

PEAT POTS. Plantable pots of pressed peat moss come in several sizes; round or square; individually or in strips of 6 or 12.

PEAT CUBES. Plantable cubes of peat moss are both pot and soil. Some contain nutrients, too.

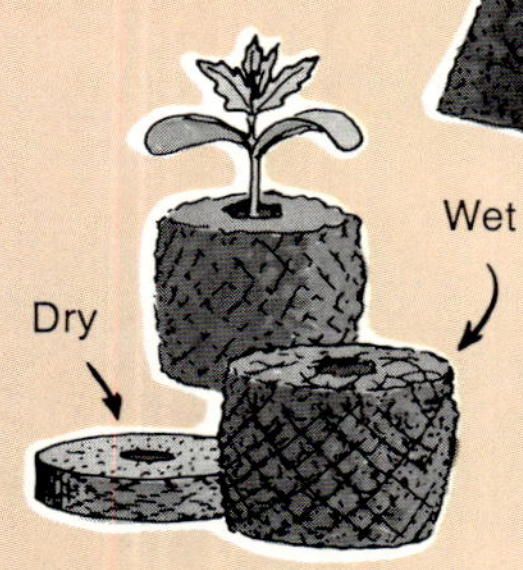

JIFFY-7 PELLETS. Improved peat cubes. Small and dry until ready to use; expand to 2" when dampened. They're reinforced with plastic netting.

MILK CARTONS. Quart or half-gallons when cut about 3" high, make excellent pots for seed starting. Tear the carton from soil ball when transplanting to avoid disturbing the roots.

SEED STARTING KITS. All the equipment you need to start seeds is available in kit form from several manufacturers. Some include a warming cable or plastic greenhouse.

The many ways to start from seed

The type of containers to hold the seed can be anything that will hold soil. However, to get the best results follow these simple guidelines:

✔Give the seed the temperature it needs for fast germination—from 70° to 80°. See chart on page 16.

✔Use a sterilized planter mix.

✔Sow seed 11 to 12 weeks before time to set out plants—after the weather has stabilized.

✔Best to plant in biodegradable containers that can go directly in the ground, container and all. For example: Peat pots, Jiffy-7's, Kys Kubes, and other compressed peat containers.

Homemade mini greenhouse.

Move seedlings from flats to pots when first true leaves are formed.

Indoors, artificial light gives necessary light for good growth.

The one-step method

Choose containers, any of those illustrated on page 16, except flats. We've found peat cubes or Jiffy-7s are the easiest to work with.

Sow seeds by putting two in each dampened container.

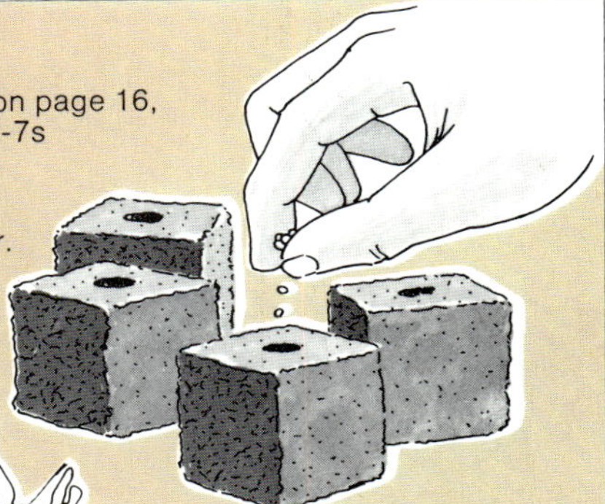

Wire Wickets

Water thoroughly and place on a tray in a large plastic bag; keep at about 75° in a bright, but not sunny, location. Keep them damp, not soaked.

When plants are 4″ to 6″ high they are ready to harden-off and transplant into the garden (see text, page 18).

The two-step method

STEP ONE
Sow seeds in a flat or shallow pan of vermiculite or other light soil mix (see text). Be sure the medium is thoroughly damp before seeding. Set seeds about ¼″ deep and in pairs about 2″ apart. Cover seeds and water lightly.

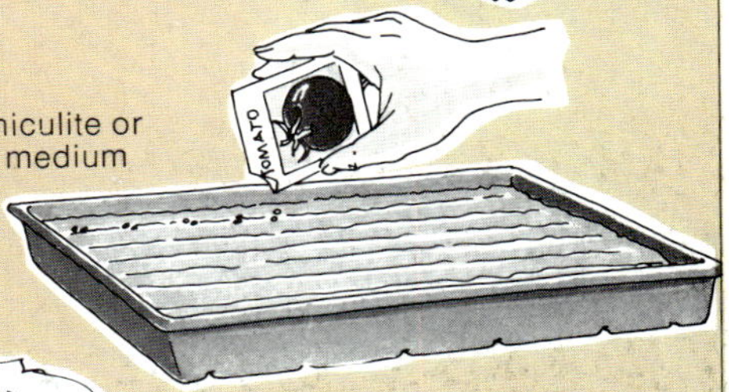

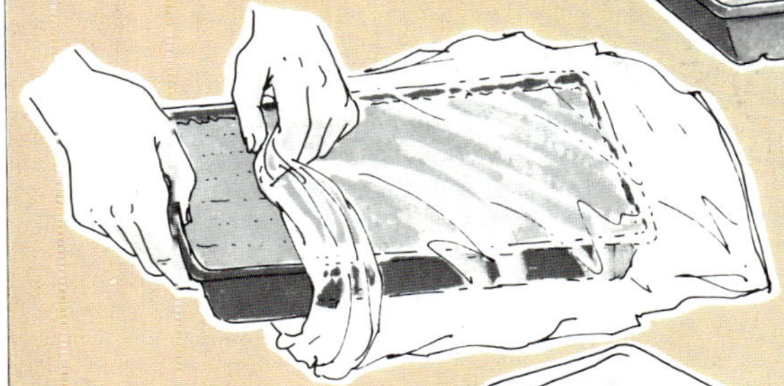

Slip the flat into a plastic bag and keep at about 75°. No additional water is necessary until after seedlings appear, and then only enough to keep the medium damp—not soaking wet.

STEP TWO
When the first true leaf is formed (it looks like a tomato leaf) the seedlings are ready to move to 3″ or 4″ peat or plastic pots filled with soil mix.

Pull the strongest seedlings from the flat and set into a hole in the soil mix so the cotyledons (seed leaves) are about ½″ above the soil.

Press the soil firmly around the roots and stem.

Wire wickets

Put the pots on a tray in a plastic bag until the plants are 4″ to 6″ high and ready for hardening-off (see text, page 18).

To keep seedlings warm and moist

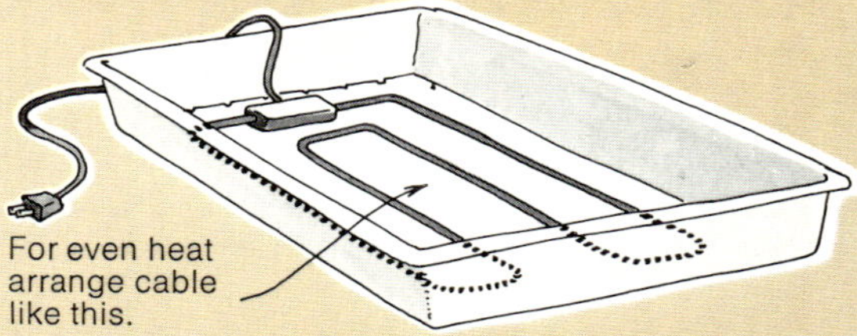

WARMING CABLE. When spread beneath the soil in a flat, it provides gentle, uniform heat for quick seed germination and optimum plant growth.

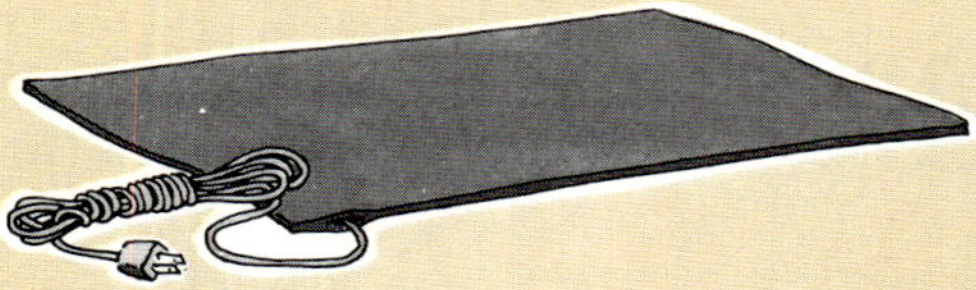

PROPAGATING MAT. A rubberized "electric blanket" to place under flats or trays of young plants to help maintain constant soil temperature.

OLD-FASHIONED COLDFRAME. In principle it's a small greenhouse; just a glass- or plastic-covered box, heated by solar radiation. It conserves this radiation absorbed by the soil for use during the night and on cool days.

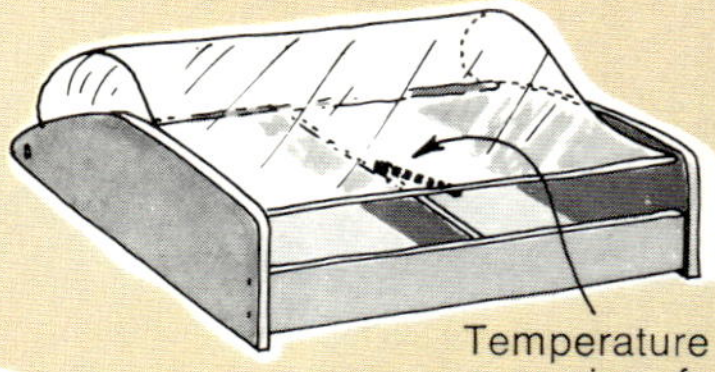

AUTOMATIC COLDFRAME. A new item on the market is this portable coldframe that opens and closes automatically when the inside reaches pre-set temperatures.

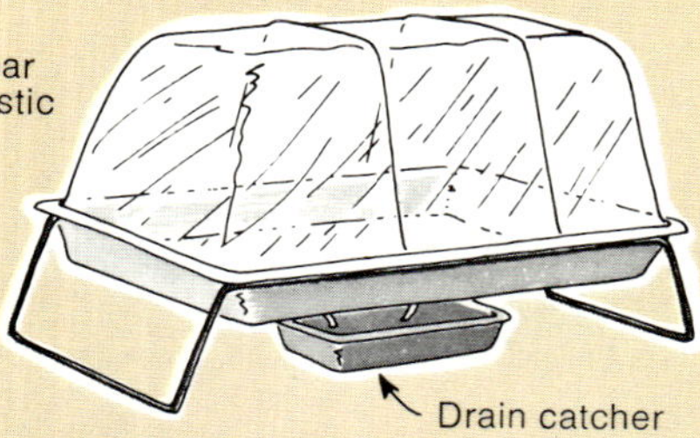

INDOOR GREENHOUSE. Many manufacturers are offering units similar to the one shown here. Some are rigid plastic, others are plastic film with wickets. Some include warming cables or built-in heating units.

Hardening-off

THE IDEAL TRANSPLANTS, when just right for hardening-off, are stocky—preferably wider than they are high; have 4 to 6 true leaves; and do not yet have flowers or flower buds.

PORTABLE LATH HOUSE, about 6" above the plants, gives the right kind of hardening-off protection from sun, wind, and cool temperatures.

A LITTLE RED WAGON or a wheelbarrow is great for hardening-off if there's still a chance of frost. Give the plants a little sun by day and move them into the garage at night.

Transplanting

It has been our experience that while the accepted advice on handling transplants is excellent in theory, it is difficult to follow in the ordinary tomato-growing practices.

We know, for example, that the ideal transplant is a plant with four to six true leaves, wider than tall, stocky, young, and succulent. This plant should not be fresh from the greenhouse but slightly "hardened" by subjecting it to outdoor conditions.

But buying that ideal or producing it at just the right time for outdoor planting is not always possible.

The bedding-plant grower has a series of target dates. He tries to deliver the ideal transplant for the early-planting customers and all those who follow through the transplant season. If the target date turns out too cold and rainy for any kind of gardening, either the bedding-plant grower or the retailer will have plants on hand that may be a week or two older before they are sold.

If the plants are in six-packs, the stems of the plants are only about 1½ inches apart. The chances of plants being wider than tall are remote. If there are delays in sale at the retail outlet, the plant will be tall and spindly. Gardeners who insist on transplants at the earliest hint of summer weather cause a lot of trouble. The bedding-plant grower and the informed nurseryman know that planting before the weather has settled is risky.

Let's say that indoors, or in your greenhouse, the young plant has grown past the ideal size for transplant and has moved into the flowering and fruit stage before the weather is right for planting in the garden. Won't you get earlier fruit production from that plant than you would from a fresh young transplant?

The answers are "yes" and "no." The "yes" answers come from growers who feed the transplants liberally throughout its stay in the greenhouse and give it the root space of a six-inch pot. The "no" answers come from those who have tested the effect of age of transplants without qualifications.

Tests showed that growing plants three to four weeks longer in the greenhouse seldom shortens the time to the mature-ripe fruiting stage by more than a few days.

The reason for getting the transplant into it's permanent location in its youthful stage is this: If you allow the transplant to flower and set fruit, and then plant it in the garden, it will not have enough roots at the time, to support the fruit. Even in a greenhouse operation, you're better off to move the ideal-sized transplant into a 5-gallon container, if you want a good yield of fruit.

An experienced tomato grower had a different answer: "Yes, you can move out an overgrown transplant if you take these precautions." He gave the transplant the root space of a six-inch pot and enough bench space for the leaves to spread. During its stay in the greenhouse the transplant was fed regularly. When the overgrown transplant went into the garden, it was fed again. There was practically no interruption of growth.

A "no" rating came from another experiment—a test report on transplanting tomatoes when the fruits were the size of marbles.

Transplants held in the greenhouse until fruits of marble size had developed were planted in the garden in this fashion: half of the plants were set out with the fruits on the vine; all of the fruits were removed from the other half.

The plants with fruit left on the vine ripened the already existing fruits—and then quit. The vines with the fruits removed continued to develop and produce a succession of crops.

The answer to the question, "Can you substitute time in the greenhouse for time outdoors in the garden?" is this:

Yes, if you transplant the seedling to a large container (5-gallon size or larger) and transfer the container plant outdoors when the weather is right. Gardeners in areas where the season is too short or too cool for the late-season varieties manage to grow the late 'Beefsteak' variety in this fashion.

So the best answer still leaves us with "yes" and "no."

Don't forget

Try to get transplants of ideal size. Ideal or otherwise, plant them deep.

How to transplant

PREPARING THE SOIL. Level and smooth your tomato patch. Spread organic matter and a good pelleted tomato food over the planting area (Be sure to follow the directions on the label. They will tell you how many pounds for each 100 square feet.) and dig it into a depth of 8 to 10 inches.

DIGGING HOLES. Mark where you want each plant and dig the hole deep enough to bury the tomato stem as far as the first leaf.

STARTER SOLUTION. Soak each hole with a cup or so of high-phosphorus starter solution about ½-hour before planting. No additional fertilizer will be needed until after the fruit sets.

READY THE PLANT. If it's in a peat pot, tear the top edge off so it can't act as a wick and dry out the root ball.

If it's in a plastic, fiber, or clay pot, tip it out—don't pull it by the stem.

If it's in a peat cube it's ready already.

PLANT IT DEEP. Put a good stocky plant in so the first leaf is just above the soil.

Leggy plants should be buried very deep. Pull off a few leaves and, to make planting easier, bend the stem and lay it in sideways . . .

Roots will develop all along the buried stem.

For comparison of deep versus shallow planting see "Experiments," page 57.

FIRM THE SOIL. Press the soil firmly around the root ball. Then water lightly to settle the soil and remove any air pockets that may be left around the root ball.

Soil

If you are blessed with a soil that is deep and fertile, easy to work, and easy to manage, any remarks we make about soil are not for you. In our tomato-growing experiences we have had to contend with several problem soils—a few have been too sandy, most have been heavy clay. It's true that the tomato plant will thrive in heavy clay soil if you water *infrequently* but *deep*. Making heavy soils more friable by adding organic matter may not make a big difference in the life of the tomato but it does make a difference in the life of the gardener.

The addition of organic matter—compost, peat moss, manure, sawdust, ground bark—makes clay soils more mellow and easier to work. Organic matter opens up tight clay soils, improves drainage, and allows air to move more readily through the soil, warming it up earlier in the spring. In light, sandy soils, organic matter holds moisture and nutrients in the root zone. The more organic matter you add to a sandy soil, the more you increase its moisture-holding capacity.

The quantity of organic matter must be large enough to physically change the structure of the soil. Enough means that at least one third of the final mix is organic matter. To add this amount, spread a layer of organic matter over the soil at least two inches thick and work it in to a depth of six inches.

You can't change soil structure beneficially with a little dab of anything. A little peat moss or a little straw and compressed clay soil makes a good adobe brick. (A word of caution: If you're adding peat moss to condition the garden soil, moisten the peat moss *before* you mix it into the soil. Wetting peat moss is not easy. We wet it by putting the amount we want to use in a plastic bag, adding *warm* water, and then squeezing and mixing the bag by hand.)

The total area of soil in which you are going to plant tomatoes will probably be relatively small. A planting of eight to twelve vines—early and mid-season varieties—will supply enough tomatoes for the average family with a fair appetite for tomatoes and an intent to can.

We use the following two planting methods:

1. If the plants are to be trained on a trellis we condition the soil in a trench about three feet wide and 14 feet long. This trench will receive seven plants. One four-cubic-foot bag of peat moss and four cubic feet of vermiculite or perlite are spread over the 3 x 14-foot area. After roughly mixing peat moss and vermiculite or perlite, two pounds of 5-10-10 dry fertilizer are sprinkled evenly over the area.

All are then mixed into the soil. We use a tiller that mixes to a depth of six to eight inches.

2. If the tomatoes are to be planted in cages, we condition the soil only in the circle in which the tomatoes will grow.

Although a tomato plant will grow in a soil that is slow to drain, too frequent watering clogs such soil and causes trouble. Gardeners with poor watering habits should grow tomatoes in a soil that can be overwatered.

Soil tests

Soils that are ideal for rhododendrons and azaleas are not the best soils for tomatoes.

For gardeners who want to grow the perfect tomato, the information provided by a soil test is essential.

The pH level and the lime requirement are by far the most important items in a soil test. The tomato plant requires a pH level of 6.0 to 7.0 for best growth.

Gardeners who live in areas where the liming of lawns is a yearly affair can be sure that their natural soil is low in calcium—an element needed by the tomato to combat blossom-end rot.

If you have had trouble with this disorder, do this: Before planting add 5 pounds of pulverized limestone (adds calcium) to 100 square feet. Mix thoroughly throughout the top 8 to 12 inches of soil.

Many states offer free or low cost soil tests. See your County Extension Agent for details.

Synthetic soils

If you are growing tomatoes in containers, we think that the use of a quality-controlled, synthetic soil mix is almost a must.

It gives the plant the soil it needs for vigorous growth and fruit production. From both the viewpoint of the tomato and your own gardening enjoyment there is much to say in favor for these lightweight "soil" mixes.

Garden stores everywhere sell special soil mixes under a wide variety of trade names: *Redi-Earth, Jiffy Mix, Metro Mix, Super Soil, Pro-Mix, Baccto Potting Soil, Terra-Lite Tomato Soil,* and many others.

Some mail-order seed companies package their own brands of potting soil and special tomato soils. For example, Burpee Seed Company offers their special tomato mix, *Burpee Tomato Growing Formula,* and Park Seed Company has their list of *Sure-Fire Mixes.*

Other seed companies may list their own mixes, or widely distributed commercial mixes.

The basic ingredients for artificial soils and why. The organic percentage of the mix may be peat moss, redwood sawdust, shavings, bark of hardwoods, fir bark, pine bark, or a combination of any two.

The mineral percentage may be vermiculite, perlite, pumice, builder's sand or granite sand, or a combination of 2 or 3 of them. The most commonly used minerals are: vermiculite, perlite, and fine sand.

Vermiculite (Terra-Lite) when mined resembles mica. Under heat treatment the mineral flakes expand with air spaces to 20 times their original thickness.

Perlite (Sponge rock) when mined is a granitelike volcanic material that, when crushed and heat-treated (1500°-2000°F), pops like popcorn and expands to 20 times its original volume.

Give the tomato plant the soil it needs

When soil is too heavy or too light, add organic matter—peat moss, ground bark, compost, and the like.

Keep the soil pH in the range of 6.0 to 7.0.

Have your soil tested. Add lime if low in calcium.

Shown are ingredients used in creating synthetic soils. Top row: Vermiculite, sand, bark. Bottom row: Perlite, peat moss, Jiffy-Mix (50% vermiculite and 50% peat moss).

The mix you buy may be 50% peat moss and 50% vermiculite, or 50% ground bark and 50% fine sand, or other combinations of the organic and mineral components. The ingredients in the mixes vary but the principle behind all mixes is the same—soilless "soil" must provide:

1. **Fast drainage of water through the "soil."**
2. **Air in the "soil" after drainage.**
3. **A reservoir of water in the "soil" after drainage.**

Most important in any container mix is *air in the soil* after drainage. Plant roots require air for growth and respiration. In a heavy garden soil, the space between soil particles (the pore space) is small. When water is applied to the soil it drives out air by filling the small pore spaces.

In an artificial soil mix, you have micropores and macropores (small and large pores). When the mix is irrigated, water is retained in the micropores but quickly drains through macropores allowing air to follow.

Container soil must have better drainage than garden soil. Water moves through a column of soil in the garden with continuous capillary action (blotter action). Break that continuity with gravel, the bottom of the container, or what have you, and water will build up where the continuity is broken. A drop of water needs another drop of water behind it to drip out of a pot or into a layer of gravel. The amount of air in the soil after drainage is the important factor in the growth of the plant.

Keep it simple. Gardeners who like to work out complicated mixes of 5 or 6 ingredients find it difficult to accept the fact that a simple combination of peat moss and vermiculite, or perlite, or fine sand, gives good results.

The quality soil mixes are ready to use as is. Fill the container (5-gallon size or larger) with the mix, set the transplant in it, and watch it grow.

Don't assume that the mix can't be tampered with. The addition of sand (10% by volume) will merely add to the weight of the mix. The mix will still be free from disease-causing organisms, insects, and weed seeds.

In 5-gallon containers, with a large vine trained on sticks or tomato towers, we like a mix that is slightly heavier than the mix that comes in the bag. Mix, with sand added, is heavy enough to prevent vine and can from toppling over in the wind.

Buy a 2-cubic foot bag and you have enough "soil" for four 5-gallon size containers.

Remember that when growing tomatoes in containers, the smaller the container the more frequently you must water and fertilize. You are compensating for the constricted root space.

Frequent watering leaches nitrogen and potassium through the soil mix in the container—which may seem like a waste of fertilizer, but when water leaches elements through the soil it also prevents a build-up of the harmful salt content of the mix. Salts in the irrigation water are leached through the soil. Water plants in containers until the water drains out of the container.

The amount of nutrients required daily is very small. But the supply must be continuous. When frequent waterings are needed in hot weather the leaching action of the water drains away a part of the nutrient supply.

That's why experienced container gardeners gear their fertilization program to the watering schedule rather than the calendar. To them, continuous feeding with a weak nutrient solution is ideal.

In containers

Use one of the quality controlled mixes.
They provide:

Fast drainage of water with good water retention.

No disease organisms.

No weed seeds.

Plant protection

The protection of young tomato plants from the vagaries of springtime weather is carried out in many ways—from the time-honored Hotkap to a parade of home-garden devices using clear plastic, recycled plastic jugs, coffee cans, and what not. Some of the inventions are illustrated in this chapter.

All devices are based on this principle: trapping the heat from the sun during the day; slowing the reradiation of that heat at night.

There is a danger of heat build-up in the use of any protection made of paper or clear plastic. When using any device, be sure you provide for air circulation. A sudden change in the weather from grey clouds to bright sunlight has cooked many a plant by building up temperatures in an unventilated covering.

The tomato measures night temperature by the number of hours of darkness when temperatures are above 45° or 55°. If a cold night is predicted, gardeners can take advantage of the *higher* temperatures of the *first* hours of darkness by slowing the reradiation of heat from the soil.

We accomplished this in our own garden by covering a raised bed, where we were starting tomatoes, with a sheet of fiberglass (see illustration), and compared hourly temperatures beneath the panel with the outdoor temperatures until midnight. Temperatures beneath the panel were from 4° to 10° higher than they were outdoors. We theorized that the spread in temperatures was due to variations in wind velocity. Tomatoes grown under the fiberglass panel were larger, better-looking plants than any grown in the open.

To the painstaking gardener, with only a dozen plants to care for, no step in plant protection is "too much trouble."

If strong winds are a daily affair, wind protection will increase the growth and the yield of the tomato plant. When the plant is under stress from wind, its growth processes are diverted to self protection, instead of the setting and bearing of fruit.

In areas where the first frost of fall is followed by a stretch of warm weather, tents of plastic are used to extend the harvest by another two or three weeks.

In the tropics where rainfall is excessive for the best growth of tomatoes and other vegetables, the roof of fiberglass panels solves the rainy season problems for the plants, and adds much to the pleasure of gardening.

Don't forget

Most tomato varieties need night temperatures (at least in the early hours of darkness) of 50°-55° F for growth and fruit set.

Trap the heat during the day.

Slow the release of heat in the first hours of darkness.

All types of day-time covers must leave air circulation to prevent heat buildup.

Frost protection for transplants

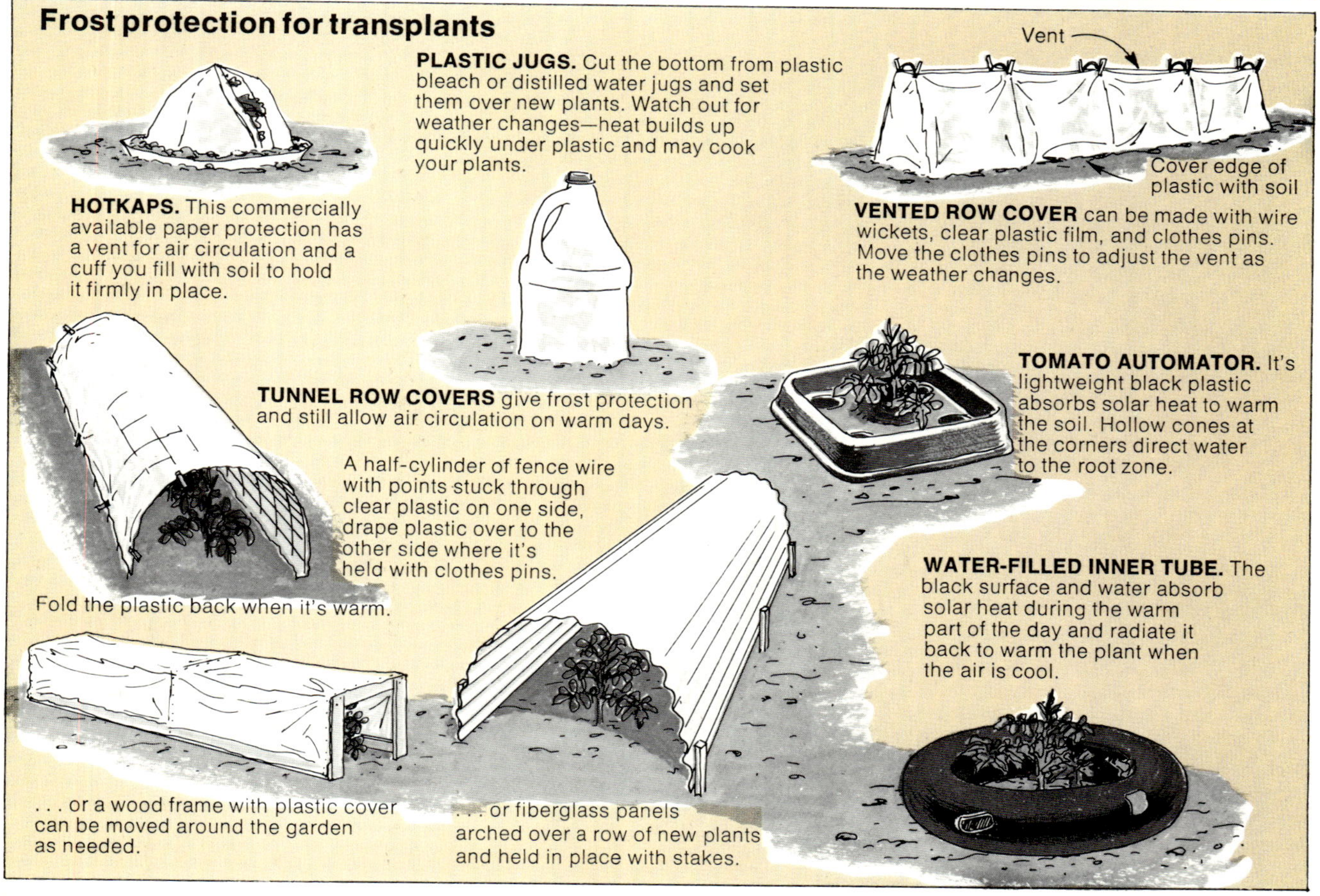

HOTKAPS. This commercially available paper protection has a vent for air circulation and a cuff you fill with soil to hold it firmly in place.

PLASTIC JUGS. Cut the bottom from plastic bleach or distilled water jugs and set them over new plants. Watch out for weather changes—heat builds up quickly under plastic and may cook your plants.

VENTED ROW COVER can be made with wire wickets, clear plastic film, and clothes pins. Move the clothes pins to adjust the vent as the weather changes.

TUNNEL ROW COVERS give frost protection and still allow air circulation on warm days.

A half-cylinder of fence wire with points stuck through clear plastic on one side, drape plastic over to the other side where it's held with clothes pins.

Fold the plastic back when it's warm.

TOMATO AUTOMATOR. It's lightweight black plastic absorbs solar heat to warm the soil. Hollow cones at the corners direct water to the root zone.

WATER-FILLED INNER TUBE. The black surface and water absorb solar heat during the warm part of the day and radiate it back to warm the plant when the air is cool.

. . . or a wood frame with plastic cover can be moved around the garden as needed.

. . . or fiberglass panels arched over a row of new plants and held in place with stakes.

Plastic fiberglass panel increases night temperatures around tomato seedlings in raised bed

Gallon-size plastic jugs with bottoms cut out are useful as mini-greenhouses.

An ordinary plastic bucket is converted into a small greenhouse with good air circulation.

It is used to protect young tomato plants growing in black plastic mulch.

Clear plastic gives increased warmth and wind protection to seedlings in wire cages.

Row of tomato plants are given wind protection by clear plastic pinned to wire support.

Feeding

Home gardeners take care of the nutrient needs of the tomato in many ways. Every type of product is used—fish fertilizer, blood meal, hoof and horn meal, manure, commercial liquid, and dry fertilizers.

Giving exact recipes for fertilizing tomatoes is like trying to write a cookbook using an oven of unknown temperature controls, flour that varies from fine to coarse, milk that may be sweet or sour, and water of unknown quality.

The first step in using fertilizer is to understand the label on the bags and packages, whether it's a labeled tomato food, or what-have-you.

All commercial fertilizers are labeled by the percentages of nitrogen (N), phosphorus (P), and potassium (K) they contain. There are many formulas:

NPK	NPK
4-12-2	5-10-10
6-20-10	16-18-6
18-24-6	18-18-21
10-10-10	and so on.

But the listings are always in the same order, with nitrogen first, phosphorus second, potassium third.
In applying fertilizers, the percentage of nitrogen in the formula dictates the amount to be applied. If the fertilizer you are using has a higher percentage of nitrogen, use less fertilizer. For example, if you have been using a 5-10-10 fertilizer and switch to one with a 10-10-10 formula, use half as much. Gardeners who do not read directions on the package may double or triple the amount that should be applied if they don't watch the nitrogen percentage.

Remember, whenever using fertilizer, don't out-guess the manufacturer. Read the label, and follow directions.

Not knowing the texture, or fertility of the soil in your garden, or the fertilizer you will use, we can't be as specific as we would like to be about a fertilization program. The best we can do is to emphasize the *differences between the special requirements* in fertilizing tomatoes.

The first step in any fertilization program for tomatoes is to give the roots of the newly set-out plant ready access to a generous supply of phosphorus. The element phosphorus does not drain through the soil like nitrogen and potassium. Therefore it must be placed where the newly formed roots can get to it.

There are three ways to satisfy the high phosphorus requirement of the newly set-out plant in garden soil.

1) Mix the fertilizer thoroughly into the soil in the area to be planted.
2) Sidedress transplants with the fertilizer at the time of planting.
3) Apply a starter solution when setting out transplants.

Use each method this way, as we have done:

1. Apply a dry fertilizer at the specified rate per foot of row (⅓ of a pound for 10 feet of row with the 5-10-10 formula), and spread the fertilizer over the area to be planted; mix into the top 6 inches of the soil.

2. Apply in narrow bands or furrows 3 to 4 inches away from the center of the transplant hole in a 3-inch-deep trench.

Careless placement of the band too close to the transplants will burn the roots. Best way is to stretch a string where the row is to be planted. With a corner of a hoe, dig a furrow 3 inches deep, 3 to 4 inches to one side of, and parallel to the string. Spread the fertilizer in the furrow and cover with soil. Repeat the banding operation on the other side of the string; then plant underneath the string.

For widely spaced plants, fertilizers can be placed in bands 6 inches long for each plant or in a circle around the plant. Run the bands 3 to 4 inches from the plant base.

3. Prepare holes large enough and deep enough to receive plants.

In a pail, prepare dilution of a high-phosphorus fertilizer according to label directions. Pour solution in each planting hole at recommended rate.

Follow-up

The tomato plant has a lot of work to do. Clusters of fruit call for abundant food supply. Nutrients and moisture must be constantly available for continuous leaf growth and fruit development.

When the first fruit is golf-ball-size start your maintenance fertilization program and continue throughout the growing season.

Container growing

Commercial synthetic soil mixes contain the necessary phosphorus for good plant growth.

Start regular fertilization program 2 to 3 weeks after setting out transplant and continue throughout the growing season.

Remember, you must compensate for the restriction in root space and the frequency in watering by more frequent fertilizer applications than when the tomato plant is growing out in the garden.

Timed-release

You get a continuous supply of nutrients from start to finish with timed-release fertilizer. Several such fertilizers are available. They vary in the nitrogen-phosphorus-potash ratio:

NPK	NPK
6-18-6	10-20-5
14-14-14	18-6-12
7-40-6	

Choose one of these ways to fertilize:

Program A
Planting in garden soil
1. Apply before planting. Use fertilizer high in phosphorous.
2. Use starter solution when setting out transplants.
3. When fruit is set—use same fertilizer as in Step 1.
4. Follow-up application once or twice a month, with same fertilizer as Steps 1 and 3.

Program B
Planting in garden soil
1. Same as Program A
2. Same as Program A
3. When fruit is set, Apply fertilizer high in nitrogen.
4. Follow-up application once or twice a month with same fertilizer as Step 3.

Program C
Containers with synthetic soil
1. Nutrients in soil mix take care of the first 2 to 3 weeks.
2. Start regular feeding program making applications according to size of container and frequency of watering. For example, a 5-gallon container would be fertilized more frequently than a 10-gallon container.

Program D
Timed Release—garden or container
1. Mix Timed Release fertilizer with soil when setting out transplants. Follow manufacturers label instructions.
2. This single application takes care of the plant throughout the growing season.

Watering

Frequency of water supply depends upon the stage of growth of the vines, the daily temperatures, the humidity, air movement, light intensity and the type of soil.

In one commercial planting in an area of no summer rainfall where summer temperatures average 80°F, tomatoes growing in deep clay loam are watered only twice a month.

In another planting, where rain supplies a portion of the watering needs, irrigation is scheduled on a "when needed" basis.

Commercial plantings are generally on deep soil where the deep-rooted tomato vine finds an adequate reservoir of water. The soil in the surburban garden may or may not be on "tomato" land. If soils are shallow or sandy, twice-a-week watering may be necessary.

If you garden in heavy soil, let the irrigation be as infrequent as the plant will allow. Don't keep the air spaces in the soil filled with water.

Regardless of how you water remember that the tomato plant will get into trouble if its water supply is sporadic. The most frequent cause of blossom-end rot is an uneven water supply. See chapter on *Trouble-Shooting,* page 58.

An effective aid in maintaining an even, continuous supply of moisture is a mulch—organic or black plastic. See *Mulches,* page 26.

Container watering

Watering a four-foot-high tomato plant in a six-inch-deep box by hand is practical if you are home in the growing season and enjoy watering. If hand-watering becomes a chore, set up a multiple-container watering system with one of the drip irrigation units.

Remember

Whether you're growing tomatoes in containers or in garden soil don't forget your objective in watering is to provide a continuous supply of moisture. Keep the following points in mind:

• When filling containers with soil, be sure to leave enough room for water: 2 inches for the standard 5-gallon container; 3 inches for a 10-gallon container.

• Water container thoroughly—until the water runs out the drainage hole.

• To prevent root rot caused by overly wet soil, don't place containers in saucers where the water accumulates.

Check out improvements in automatic watering systems. Manufacturers of drip, trickle, and ooze equipment are developing foolproof systems for the home garden. New filters, new pressure regulators, new methods of metering are becoming available.

Simple ooze system meters water uniformly and miserly throughout the planting area.

A low cost, easy to install emitter. It can be installed directly into the main line or taken to the plant by spaghetti hose.

A compensating emitter (1 gal. per hr.) that can maintain an even flow rate on non-level terrain.

The small "spaghetti type" hose can be adapted to most situations either hanging or on the ground.

A must for any drip system is a filtration system. Simple line filter above is one example of what is available.

Mulches

Each type of mulch has a different purpose. Both the clear plastic and the black plastic warm the soil. They are effective early in the season. In cool summer areas, where an increase of soil temperature is welcomed, the clear plastic is used.

The black plastic with only a slight warming effect has the advantage of stopping all weed growth.

The reflective mulches—aluminum foil and mineral-coated plastic—and the organic mulches, are used to reduce soil temperatures and should be applied *after* the soil has warmed up.

Reflective mulches have the added benefit of increasing the light on the lower leaves that normally are shaded out by the foliage above them.

Our favorite mulch is the black plastic film. A black polythylene mulch does a good job of keeping tomatoes off the ground and delivers yields equal to that of any type of training. Field tests indicate that high fruit quality is possible when plants are allowed to sprawl naturally over the plastic. Strong-growing tomato plants should be given a space of three by five feet when grown on the plastic.

Temperature readings of the soil beneath black plastic show that the increase is generally in the 3° to 6°F range, sometimes only 2°F. The temperature of the film itself soars high on hot sunny days and kicks back a great deal of heat to the air above it, rather than transferring it to the soil.

Black polyethylene is available in 1 to 1½ mils-thickness, in rolls three to six feet wide. Get the 1½-mils thickness.

Florida test

A mulch of black plastic increased yields in a test planting by the University of Florida in commercial growing fields. The raised beds were 6 inches high, 48 inches wide, with 16 inches between the rows. After the beds were formed, fertilizer was mixed thoroughly into the 6-inch depth of the bed, and the beds covered with a black plastic mulch. Two rows of tomatoes were grown on each bed. This system is also successful with cucumbers, eggplant, peppers, squash, strawberries, and other crops.

Organic mulches

In the booklet *Commercial Production of Greenhouse Tomatoes,* Agricultural Handbook No. 382, the U.S.D.A. has this to say about mulching of greenhouse tomatoes (the same practice of the application of organic mulches can be followed in the garden):

"A common practice in greenhouse tomato production is to use an organic mulch. Mulches serve the following purposes: (1) To help maintain more uniform soil moisture, thereby lessening fruit cracking; (2) to add organic matter to the soil; (3) to reduce soil compaction and thereby improve soil aeration and root growth; (4) to protect the fruit of the first cluster from touching the soil and becoming infected with soil rots; (5) to reduce the amount of spattering of water drops from the soil onto the lower leaves, thereby reducing the spread of diseases.

"A number of materials are satisfactory for mulching greenhouse tomatoes. Some of the most commonly used mulching materials are wheat straw, peanut hulls, ground corncobs, and manure. Other materials used include alfalfa, red clover, wheat and rye straw, various types of hay, sawdust, wood chips, pine needles, sphagnum peat, rice hulls, and sugarcane bagasse.

"The choice of mulch depends on the cost and availability of the material, its ability to prevent soil compaction, its decomposition rate, its effect on the nutritional balance of the soil, its freedom from potentially harmful chemicals such as herbicides, and the possibility that it may carry a disease-producing organism into the greenhouse.

"The mulch may be applied after the soil has been sterilized and cultivated just before planting, or it may be applied after the transplants have been set in the ground beds. The mulch should be spread over the entire soil surface to a depth of 2 to 3 inches."

Black plastic

In one of our test gardens we have planted 52 varieties of tomatoes, a total of several hundred plants. Half of them are mulched with black plastic. These plants are better looking, greener, and more succulent than those handled in any other fashion. Two irrigation systems have been installed. One bed is watered by the ooze system using Viaflow. The other system is trickle, or drip irrigation.

A mulch of black plastic or an organic mulch evens the water supply to the plant by reducing the water loss through evaporation. There is no drying out of the top inch or so of the top layer of soil. The loss of moisture by the leaves of the plant (transpiration) increases as the plant grows and forms additional foliage. The loss of water through transpiration of a vigorously growing vine is far greater than the loss through evaporation. Gear your watering schedule to the needs of the plant. Don't expect the black plastic to take care of your watering job.

Reflective mulches such as aluminum foil have their place in tomato growing in containers as well as in garden soil. See *Experiments,* pages 54-57.

This season we are using circles of black plastic in 5-gallon cans. We may be fooling ourselves, but the vines seem healthier and all moisture tests are favorable. Using the mulch, we get root action in the top inch of soil.

The automator

The use of the automator with a black plastic mulch (see photograph, opposite page) benefits the tomato grower in many ways:

✔It eliminates the need for a watering system beneath the plastic.

✔Combining it it with the black plastic increases the soil area protected from weed growth and water loss through evaporation.

✔It gives the gardener a way to add water and fertilizer, accurately and deeply, through the black plastic.

Watering and plastic mulches

When using plastic film (or aluminum foil), consider how you water the tomato plants *before* applying the plastic mulch.

We have used these methods of watering:

1. Laid down Viaflow, at center of bed.
2. Installed drip irrigation at 24-inch intervals.
3. Cut slits (inverted "T") in plastic to allow water (rain or from the hose) to penetrate.
4. Used combination of "automator" and black plastic.

Left: Shallow trench is dug on each side of the row, the approximate width of the plastic. Plastic is unrolled the length of the row.
Right: Plastic is unfolded evenly along the row; edges are secured by burying them in the small trenches and covering with soil.

Drip irrigation system supplies water beneath black plastic mulch. Install before plastic is laid down.

Automator combined with black plastic mulch makes watering and fertilizing plants a simple chore.

Aluminum foil mulch cools soil and reflects additional sunlight back into plants.

Aluminum foil mulch, cut to fit, is placed beneath tomatoes trained in wire cages.

How many stems?

ONE STEM SYSTEM. If getting tomatoes a couple of weeks early is important to you, training to a single stem will usually do it. Just pinch out each new stem after it's sprouted at least two leaves as shown below.

Be aware, however, this severe pruning will reduce your total crop greatly and also is likely to increase the incidence of blossom-end rot (see page 60).

TWO OR MORE STEMS produce more tomatoes with better foliage protection from the sun. Choose the stems you want to keep and pinch out the others as they develop. Again, be sure to leave a couple of leaves at the base of each stem you remove.

The 4-stem plant shown here is off to a good start. But it, as well as the plant shown above, will need support. The following pages will show you how.

Where to pinch-out unwanted growth

Before removing suckers or side shoots on a tomato plant, wait until two leaves develop and pinch above them. This practice provides better foliage cover to protect the fruit and stems from sun damage.

Training tomatoes

That there are as many ways to train tomatoes as there are gardens, is a bit of truth repeated many times.

What is the best method?

We passed that question on to a fellow worker who has trained tomatoes for more than 50 years. We quote him:

"The training method I grew up with was the simple single-stake method. I drove 8-foot long 2 x 2's into the ground about 2 feet apart. This was for the fall-growing, indeterminate varieties. The low-growing bush type plants were allowed to sprawl.

"At the base, and on the south side, of each stake the varieties of my choice were planted.

"As the plants grew my job was to break off the sideshoots—the suckers that formed at the axil of the leaf branch and the stem. As the plant grew, the single stem, with all side branches removed, was tied at intervals to the stake."

Wire cages. "The wire cage of concrete reinforcing wire or hog-wire made a big change in my training techniques. The mesh must be large enough to allow a reach-through for picking ripe tomatoes. In a way it's all too simple. Fit the cage to the plant. Let the plant grow in its own disorderly way.

Once a pruner, always a pruner. "Regardless of all the evidence that a plant in a cage will out-yield a plant that is pruned, I can't resist some pruning. When a branch strays out of the cage I lop it off. When growing a plant that spreads wide and grows tall I feel it can spare a branch or two."

String and "sky hook" training. See art.

Other structures for support

SINGLE STAKE. The simplest way, but the vine must be tied to the stake.

LADDER TRELLIS. No need to tie if you weave the vine through the trellis as it grows.

TOMATO TOWER. This commercially made gadget supports the vine without tying. It's 5 feet tall, made of heavy galvanized wire with a 3-prong base that anchors it securely in the ground.

TEPEE. A compact group of 3 or 4 plants are supported nicely when tied to a tepee of 1-inch stakes.

WIRE FENCE on each side of a row of plants lets you slide slats through to support vines when and where needed without tying.

WIRE MESH stretched horizontally on 1-foot posts holds sprawling vines off the ground.

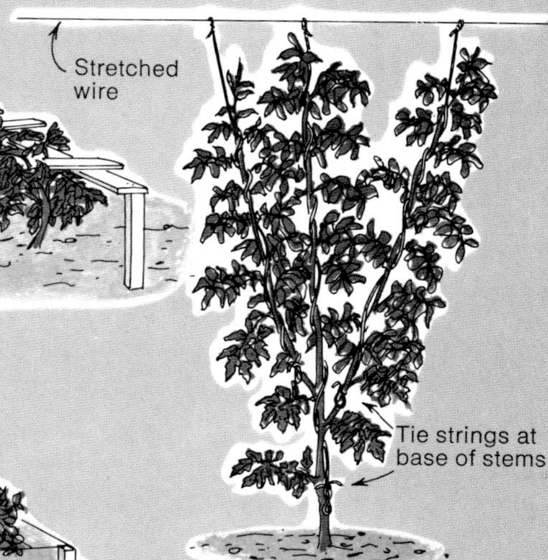

SKY HOOK. Strings from an overhead wire or beam are tied to the base of each stem. Tomatoes won't coil around the string as beans do, but the effect is the same if you help a little every few days or so as the stems grow (see below)

Attaching vines to structures

STRING OR TWINE. Use a loose figure-8 or, better yet, tie one knot tightly on the stake and another loosely on the vine.

TWISTIES. Plastic-coated wire holds securely when twisted tightly on the stake; loosely on the vine.

PLASTIC TAPE. Tree-tie tape won't cut or girdle because it stretches with growth. But, tie it loosely on tomato vines.

SELF-LOCKING GADGETS. There are several kinds of plastic ties available. Be sure you don't set them too tightly.

SKY HOOK. Here's some detail of the support method shown above . . .

Wrap the growing tip around the string as it grows.

Tie the end of the string loosely to the base of the stem.

Small cage for bush-types.

Tall circular cage for viney indeterminates.

Wire cages

When the small bush-type tomato is allowed to sprawl the weight of the ripening fruit opens up the bush and exposes the tomatoes to sunscald.

A SMALL CAGE supports the bush-types, keeps fruit and foliage off the ground, and prevents most sunscald.

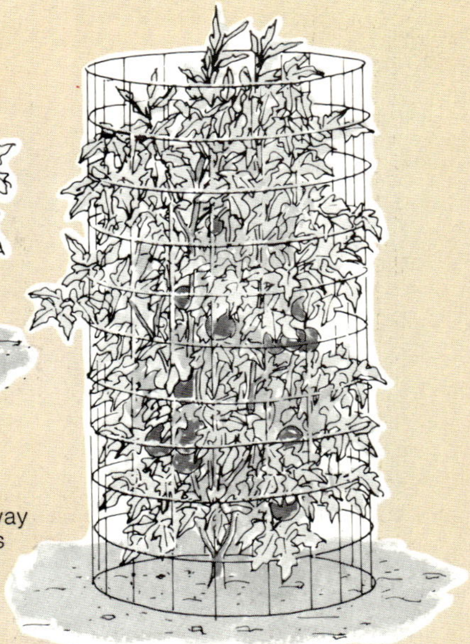

A TALL CAGE is a very good way to support viney indeterminates or strong determinates without tying. Keep the plant pinched inside the cage until it grows up and over the top.

What size wire cage?

Several years of experience in the garden, along with a recent scientific experiment (see text), have indicated the ideal size for a tomato cage. It is 24 inches in diameter and 60 inches high.

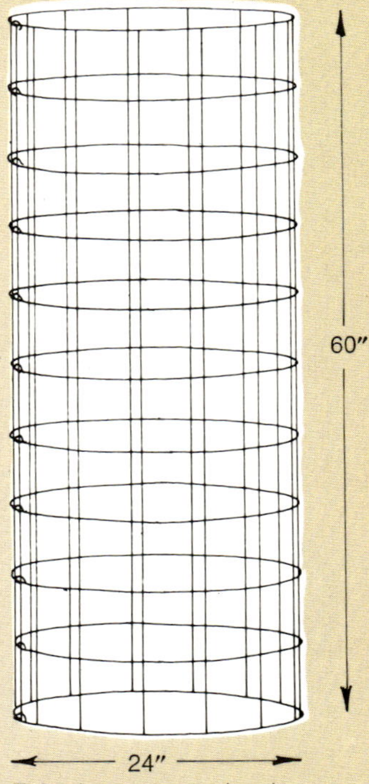

This size supports the vine very well, virtually prevents sun scald, and allows enough leaf surface for maximum fruiting and maximum solids within the fruit.

The wire cage

In our training programs we have used wire as a circular cage, as a trellis and as a double trellis. All three perform the important function of keeping the fruit off the ground.

The vine, allowed to grow in its own disorderly fashion, in the wire cage, has better foliage cover, and more protection against sun scald than when allowed to sprawl, or trained on a stake or trellis. And, you avoid the chore of tying and pruning.

The first years we used the cages we made them 18-inches in diameter. Performance and yields were satisfactory. This last season we compared performance in cages with diameters of 24-inches and 30-inches.

The 18-inch diameter cages were satisfactory for the small, bush-type vines. However, the more vigorous and indeterminate growers showed a definite improvement in performance and yields in the 24-inch diameter cages.

Six plants trained in double 2′ wide x 12′ long hogwire cage. Lath pieces through wire sides support vine and fruit. Easy to manage as the circular cage.

To make a cage

1. **SELECT THE WIRE.** Concrete reinforcing wire with 6-inch mesh is excellent because it gives good support and you can reach through to pick tomatoes—but, it does rust. After the cage is made paint it with rust-resistant paint.

   Galvanized fence wire will last many seasons without painting. It comes in many mesh sizes, either welded or woven. We like the woven better, even though its more expensive, because the welded joints do break occasionally.

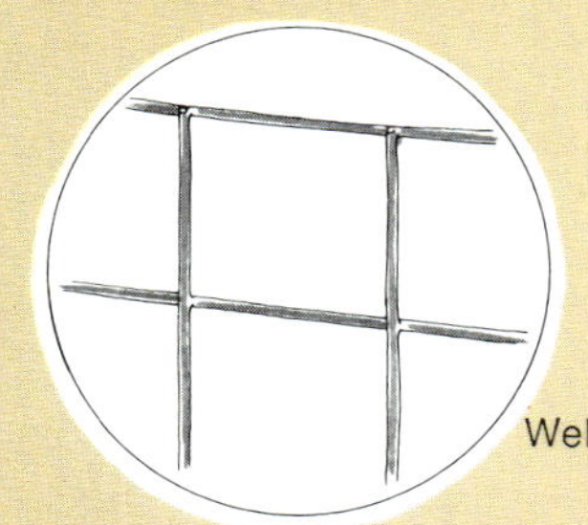

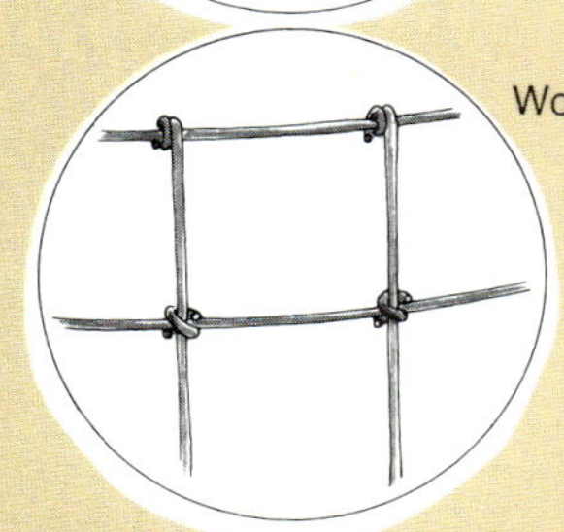

Be sure to get 4- to 6-inch mesh so your hand will fit through for harvesting.

2. **CUT WIRE TO SIZE.** Decide what size you want your cage—remember the "ideal" is 24 inches x 60 inches. Multiply the diameter by pi (3.14) to get the circumference:
   $24 \times 3.14 = 75.36$
   Make it 76 or 77 inches to allow for a little overlap and crimping.

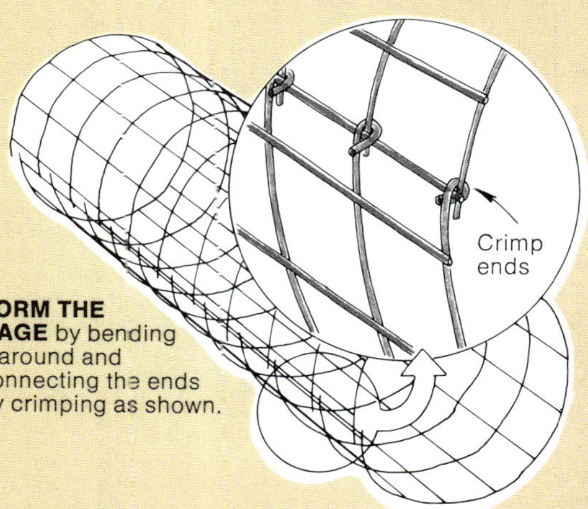

3. **FORM THE CAGE** by bending it around and connecting the ends by crimping as shown.

4. **SECURE IT TO THE GROUND** with stakes or wire wickets or remove the bottom cross-wires and push it into the ground.

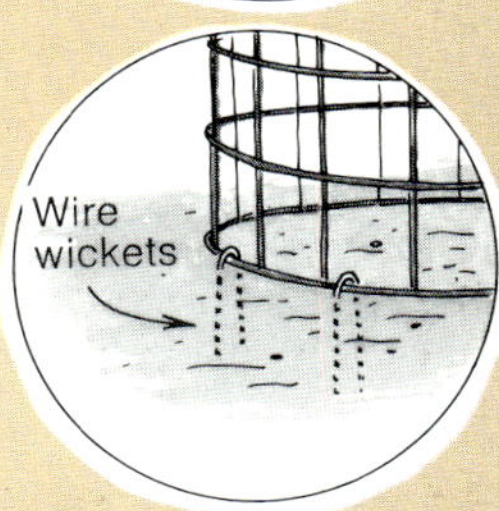

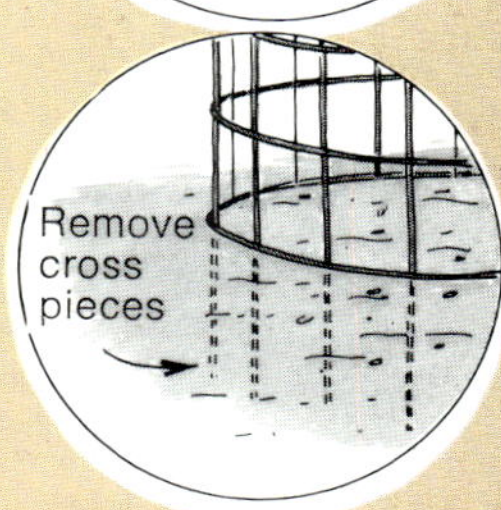

A specially designed tomato box holding 7½ gallons of soil.

Small, determinate bush-type varieties are handsome up-close in containers.

Large permanent-type container outside kitchen door.

A series of 5-gallon containers individually staked.

Plants grouped in a large container create a wall of tomatoes.

Tomato vines of two 5-gallon containers are trained on a one-legged A-frame.

Containers and hanging baskets

The practice of growing tomatoes in containers is favored by gardeners with two kinds of soil problems. Those who have an unsuitable garden soil and those gardeners who have no garden soil.

On balconies, decks, and paved patios the container route is the only way to go. Successful container growing on deck or balcony gardens depends first on the hours of sunlight the tomato plant will receive. The accepted minimum requirement is 8 hours. In our trials the 8-hour requirement was not necessarily a continuous exposure, but could be 4 hours in the morning and 4 hours in the late afternoon.

Determined container gardeners, handicapped with a difficult sun exposure, have put their containers on wheels so that the plants can be switched from one side of the balcony to the other to follow the sun.

What size of container? The minimum size depends more upon your ability to meet the plant's needs for a continuous supply of moisture and nutrients, than the size of the container.

A check of the container photographs throughout this book is the best evidence that the tomato plant will adapt itself to almost any size container.

The effect of container size on yield and plant growth are detailed in the chapter on *Experiments*, pages 54-57.

In the chapter on varieties, pages 34-49, you'll find a list of the varieties well adapted to growth in containers and hanging baskets.

Of these varieties our favorite for use in hanging baskets is 'Small Fry' and 'Sugar Lump.' 'Small Fry' has been a consistent performer year after year. The first *ripe* fruit ripens early and the vine continues to produce throughout the growing season.

As with all tomato plants the early growth is upright. To speed up the downward growth we attach small weights to the branches. The weights weigh about 1 ounce and are made up of materials on hand—nuts, bolts, lead sinkers, or steel washers.

The variety 'Red Cherry' is also a long-season producer in hanging baskets. High yields of small fruit are possible when grown indoors in late winter and early spring with plenty of light from a large window.

The 'Yellow Pear' is too rank a grower to be confined to a hanging basket.

A hanging cylinder supports the variety 'Small Fry' which produces a profusion of 1" fruit both early and late.

Hanging pots of 'Cherry' tomatoes as seen from inside house.

Varieties—their profiles and portraits

Everyone agrees there are more than enough tomato varieties. But as long as there are climate differences, local preferences, and competition among tomato breeders and seed companies, the list will be long. Here is basic information on more than 90 varieties that are tested and generally distributed.

Varieties are presented in this fashion:

A. They are listed in order of ripening.

B. Basic facts are given in a panel introducing the variety.

C. Origin and supplementary facts follow in copy.

The panel tells you this:

1	2	3	4	5	6
	**Redpak**	**VF**	**Hy**	**71 days**	**Det.**

1. Shape. Tomatoes, except paste tomatoes, are classified in these basic shapes (for shape of paste tomatoes see page 43):

The illustrations of the shape do not change by size. The size is noted in the copy.

2. Name of variety.

3. Disease resistance is shown by "V"—verticillium wilt, "F"—fusarium wilt, "N"—nematodes.

4. Hybrid variety (Hy) and standard variety (S).

5. Number of days from transplanting to first picking (71 days). This is a relative figure. It will vary by the weather, season, and when transplanted.

6. Determinate—bush type (Det). or **Indeterminate**—tall-growing type (Ind.)

◁

Top: Tomatoes vary in shape. From left to right: Globe, deep globe, oblate, and deep oblate.
Center: They vary in interior structure—the ratio of meat to gel.
Bottom: And they vary in size, measured in inches and ounces.

Availability

Varieties are available as seeds from the mail-order seed companies, as small plants at garden stores, in seed packets in seed display racks, and from wholesale seed companies.

On page 96 you will find a list of the seed companies. Each has been given a number. The catalog of each seed company has been searched for variety listings. If the variety is listed in more than eight catalogs, that variety is noted as "widely available." If the variety is available in less than eight catalogs, the number of the seed company carrying the item is listed following the description of the variety. For example:

'Redpak,' Available: (13).

By checking the catalog listings on page 96, you will find that (13) is the number of the Joseph Harris Co.

In the copy that follows the panel we have attempted to give clues that would help set the variety apart from scores of similar varieties.

We have endorsed a number of varieties which we believe worthy of a trial in your garden. Our judgements are based on reports from trials in various areas, and the variety recommendations of the Agricultural Extension Service.

A variety is described as "widely adapted" when its performance rates high in many varied climates. Gardeners in climates too cold or too hot, or with a short growing season, will favor first the local variety recommendation, and then the widely adapted varieties. Local recommendations are set apart under the heading of "Sectional Specials" in the list of varieties that follows.

As the camera sees them

In the variety list we consider tomatoes through the words of many writers. At the same time we look at many tomatoes with the eyes of the camera.

The camera gives definition to such fine words as ". . . thick-walled, firm and meaty," ". . . heavy yield of large fruit in clusters," ". . . smooth and solid."

The camera makes no attempt to glorify the tomato. We hoped that it would be objective—to tell it like it is. But we must admit that our photographers were in love with the tomato.

Missing in the eyes of both the writers and the camera is the taste quality of the variety. Only you can measure that.

Have we played fair with the tomato?

Have we failed to single out the variety that has performed so well for you? Look at any listing of tomato varieties as an invitation to explore. Set up a trial of your own—mix the new with the old.

Early-season varieties

Early Girl	V	Hy	62 days	Ind.

Origin: Geo. Ball Co. Bears early and late. High yields of firm, smooth, meaty, medium-small (4 to 5 oz.) fruits. Resistant to cracking. Sturdy vine with good foliage protection against sunscald. Stake, or cage.

Plants available at nurseries. Seed available: (5), (30).

■ *Rugged hybrid of wide adaptability. Long season indeterminate.*

Burpee's Big Early		Hy	62 days	Ind.

Origin: Burpee Seed Co. Noteworthy for production of large, thick-walled fruit. An early-ripening variety. Vine vigorous with heavy foliage. Takes trellis, or cage.

Available (5), (27).

■ *Early-ripening, rugged hybrid. Long season indeterminate.*

Sweet 100		Hy	62 days	Ind.

Origin: Goldsmith Seeds, Inc. Very large, multiple-branched clusters of very sweet 1-inch fruits with 20 or more fruit per cluster. Vine is vigorous, with narrow leaves. Recommended pruning to one stem and staking or caging. Available: (5), (18), (21), (28), (36), (41).

■ *A very unusual, worthwhile, new introduction.*

Early Boy		Hy	65 days	Det.

Origin: Dessert Seed Co. Deep scarlet, smooth-skinned, large (7 to 9 oz.) fruit. Firm flesh, thick walls, free from white core. Strong-growing determinate with dark green foliage.

Available: (2).

Spring Giant	VF	Hy	65 days	Det.

Origin: Dessert Seed Co. Large, smooth (8 oz.) fruit ripens in a concentrated harvest season. Thick walls with a small core. High-yielding, 1967 All-America Selection. (First tomato to win this award.) Poor sunscald protection unless grown in a short cage. Widely recommended.

Widely available.

■ *Rugged hybrid with good disease resistance.*

Springset	VF	Hy	65 days	Det.

Origin: Petoseed Co. Medium-sized (5 to 6 oz.), firm fruit on vigorous, open vine. High-yielding, concentrated harvest. Good crack resistance. Poor sunscald protection unless grown in a cage or on a short trellis.

Widely available.

■ *One of the most widely recommended and widely adapted varieties.*

Midseason varieties

Fantastic		Hy	70 days	Ind.

Origin: Petoseed Co. There are two 'Fantastics.' The 'Fantastic' without disease resistance has been around for some time. Both are high yielders of smooth, medium-large (8 oz.) fruit. Similar to 'Moreton Hybrid.' Wide adaptability, widely available. Listed by more than 18 seed companies. Produces until frost.

■*Rugged hybrid of wide adaptation. Widely available.*

Super Fantastic	VF	Hy	70 days	Ind.

'Super Fantastic VF' differs from 'Fantastic' in several ways. The shape of the fruit is deep oblate to round. It has disease resistance. Like 'Fantastic' it produces heavy yields until frost. A Geo. Ball Co. introduction and is available as a nursery transplant.

Originator (53).

■ *Same climate adaptation as 'Fantastic' but with disease resistance.*

Terrific	VFN	Hy	70 days	Ind.

Origin: Petoseed Co. Large fruits (8 to 10 oz.) are smooth, firm, and meaty. Strong vine, good foliage cover. Produces over a long season. Good cracking resistance. Cage or trellis this strong-growing indeterminate.

Widely available.

■ *Rugged hybrid with triple disease resistance. Widely adapted.*

Redpak	VF	Hy	71 days	Det.

Introduction by Harris Seed Co. Adaptation not fully tested but good reports are in from many areas—NY, CT, OR, TN. Dwarf, compact vine bears large, unusually firm fruits—practically solid flesh. Holds well on the vine. When mature the leaves have a tendency to curl, but this has no effect on yield or quality. Resistant to cracking.

Available: (13).

■ *Rugged hybrid. One of the most promising of the new introductions.*

The early ripening 'Sweet 100' bears multiple clusters of cherry-sized fruits.

	Better Boy	VFN	Hy	72 days	Ind.

Origin: Geo. Ball Co. Widely adapted and widely recommended as main crop variety. Very large (12 to 16 oz.), uniform fruit on vigorous, sturdy plants. Very heavy cropper. Just right size for slicing. Give it a continuous supply of water to avoid blossom-end rot. Stake, cage, or trellis.

Widely available.

■ *Rugged hybrid with triple disease resistance. Recommended by Agricultural Extension Services in GA, MD, NC, OK, SC, TN, VA.*

	Burpee's VF	VF	Hy	72 days	Ind.

Origin: Burpee Seed Co. One of the most widely adapted varieties and widely recommended. Firm, meaty, medium-large-size (8 oz.) fruit. Strong, sturdy vine, with heavy foliage cover. Resistant to cracking and catfacing. Grow in cage or stake or trellis.

Available: (5), (27), (50), (51).

■ *Rugged hybrid of wide adaptation. The main season favorite of all the Burpee varieties.*

	Jet Star	VF	Hy	72 days	Ind.

Origin: Harris Seed Co. An important variety in the second early season. Widely adapted. On the recommended list of more than 15 State Universities. Medium-large (8 oz.) firm fruits on compact grower with good foliage cover. Sets well. Good production. Resistance to cracking. Stake, cage or trellis.

Available: (13).

■ *Rugged hybrid that is widely adapted and widely recommended.*

	Atkinson	FN	S	75-80 days	Ind.

Origin: Auburn University. Nematode-resistant. Moderate resistance to early blight and gray leaf spot. Firm and meaty medium-sized (6 oz.) fruit. Strong grower with good foliage protection. Stake, trellis or cage.

Available: (15), (22), (24).

■ *One of the first varieties with resistance to nematodes.*

	Big Set	VFN	Hy	75 days	Det.

Origin: Petoseed Co. Sets well under both low and high temperatures. Large (8 to 9 oz.) fruits are smooth, firm, and meaty. Vine vigorous and compact. Resistance to cracking, catfacing, and blossom-end rot. Cage, or let sprawl.

Available: (4), (12), (22).

■ *Very rugged hybrid and widely adapted. Good performer.*

	Bonus	VFN	Hy	75 days	Det.

Origin: Petoseed Co. Vigorous. Fruits are medium-large (8 oz.), smooth and firm. Resistance to cracking and catfacing. It appears to have some ability to set under relatively high temperatures. Definite bush type. Grow it on short trellis, in cage, or as a bush.

Available: (15), (21), (28), (37).

■ *Rugged hybrid with triple disease resistance.*

	Floradel	F	S	75-85 days	Ind.

'Floradel'—a consistent producer of uniform tomatoes borne in clusters, both in the field and in greenhouses.

Origin: Florida A.E.S. Commercial and home-garden variety. Also popular for greenhouse culture. Sets well at low temperature. Resistance to foliage diseases. Smooth, medium-firm, thick-walled, medium large (8 oz.) fruit. Vine is vigorous with good cover. Train on trellis, or stake, or grow in cage.

Available: (22), (28), (30), (37).

■ ***Southern favorite with resistance to fusarium wilt, gray leaf spot, and leafmold. Recommended by Agricultural Extension Services in AR, FL, LA, OK, SC, VA.***

	Heinz 1350	VF	S	75 days	Det.

Origin: H. J. Heinz Co. Widely recommended and widely available. Uniform, medium-sized (6 oz.) fruit. Strong compact, determinate vine can be close planted. Good crack resistance.

Widely available.

■ ***Recommended by Agricultural Extension Services in AR, OK, VA.***

	Jim Dandy		Hy	75 days	Ind.

Origin: Ferry-Morse Seed Co. Large red, solid fruit produced on vigorous vines which set continuously and hold their size all season. Good for slicing and canning. Stake or cage.

Write to originator *(55) for availability.

	Manapal	F	S	75 days	Ind.

Origin: Florida A.E.S. Bred for humid growing conditions. Has resistance to fusarium wilt and many foliage diseases. Rated as one of the best home-garden varieties in the South it has proved itself in trials in TN, GA, NC, NJ, NB, IL, CO, ID. Highly productive greenhouse variety throughout the South and Southwest.

Available: (28), (30), (37).

■ ***Southern favorite with resistance to fusarium wilt, gray leaf spot and leafmold. Recommended by Agricultural Extension Services in GA, KY, MS, NC, TN, VA.***

	Marglobe	F	S	75 days	Det.

Grand old name. Long-time favorite. Fruit is medium-sized (6 oz.), smooth, firm, and thick-walled. Vine growth is uniform, vigorous, with good, heavy cover.
Widely available.

	Monte Carlo	VFN	Hy	75 days	Ind.

Origin: Petoseed Co. Widely available and seems to be widely adapted. Produces large (9 oz.) smooth fruits over a long season. Tall-growing, strong vine with good foliage cover. Triple disease resistance, and resistant to cracking, catfacing, sunscald, and blossom-end rot.

Widely available.

■ ***Rugged hybrid with triple disease resistance. Good reports from as widely differing climates as GA, OR, CT, PA, SC.***

	Tropi-Red	VF	S	75-80 days	Det.

Origin: Florida AES. Deep red, mild-flavored, large (7 to 9 oz.) fruit. Smooth, thick-walled, with firm interior. Less subject to cracking than many varieties grown in warm, humid climates. Compact, bush-type vine with good disease resistance.

Plants available at nurseries.

■ ***A Florida special.***

	Walter	F	S	75 days	Det.

Origin: Florida A.E.S. Good garden performance but must be harvested ripe. Resistance to leaf diseases, cracking, catfacing and blossom-end rot. Firm, medium-sized (7 oz.) fruits on compact plant, good cover. Stake or let sprawl.

Available: (2), (28), (37).

▪ *Recommended by Agricultural Extension Services in GA, MD, TN.*

	Whopper	VFN	Hy	75 days	Ind.

A Park Seed Co. exclusive with wide adaptation. Heavy yield of large (10 oz.) fruit, smooth, firm, and meaty. Vigorous, tall-growing vine with good foliage cover. Good resistance to cracking and blossom-end rot.

Available: (21).

▪ *Rugged hybrid with triple disease resistance. Widely adapted.*

	Burpee's Big Boy		Hy	78 days	Ind.

Origin: Burpee Seed Co. Probably the number-one name among tomato varieties. Long-time favorite. Very large (12+ oz.) fruit is firm, smooth, and thick-walled. Produces continuously until frost.

Widely available.

▪ *A rugged hybrid. Widely adapted.*

	Burpee's Big Girl	VF	Hy	78 days	Ind.

New introduction by Burpee. Has similar size and shape of 'Big Boy' but with VF disease resistance added. Just as 'Big Boy' acceptance was followed with 'Better Boy,' 'Wonder Boy,' 'Ultra Boy,' 'Golden Boy,' and other 'Boys' we can expect a 'Girl' series of names.

Available: (5).

▪ *A rugged hybrid of wide adaptation. New introduction with all the good qualities of 'Big Boy' plus disease resistance.*

	Marion	F	S	79 days	Ind.

Origin: Clemson A.E.S. One of the best home-garden stake varieties in the South. Medium-sized (6 oz.) firm fruit. Strong and vigorous vine with good cover. Resistance to cracking. Stake, cage, or trellis.

Available: (15), (21), (22), (28).

▪ *Southern favorite with disease resistance to fusarium wilt, leafmold and gray leaf spot. Recommended by Agricultural Extension Services in GA, MS, SC, VA.*

Three boys and a girl

Late-season varieties

	Ramapo	VF	Hy	80 days	Ind.

Origin: Rutgers University. Sets well under adverse conditions. Thick-walled, medium-large (8 to 9 oz.), bright, deep crimson fruit. Strong growing, vigorous vine. Resistance to cracking, and to blossom-end rot. Stake, cage, or trellis.

Available: (5), (13), (50).

	Rutgers Hybrid F	F	Hy	80 days	Semi-det.

	Rutgers Hybrid VF	VF	Hy	80 days	Semi-det.

The story behind the many 'Rutgers' illustrates the power of a popular name. The original 'Rutgers' was introduced in 1934. The variety gained wide acceptance as a distinct type, and a quality tomato.

In 1964, Rutgers University introduced 'Rutgers Hybrid F.' It was a 'Rutgers' type with resistance to fusarium wilt. This introduction replaced the original 'Rutgers' and with all catalog listings of 'Rutgers F' this is what you get.

Since the introduction of this hybrid several seed companies have introduced 'Rutgers' with verticillium and fusarium resistance. These are 'Rutgers' type tomatoes, not the same as the original 'Rutgers.' Dr. Bernard Pollack who introduced 'Rutgers Hybrid F,' is the originator of 'Ramapo,' which is a 'Rutgers' type with both fusarium and verticillium resistance.

'Rutgers F': Widely available.

'Rutgers VF': Plants available at nurseries. Seed available: (13).

	Wonder Boy	VF	Hy	80 days	Ind.

Origin: Petoseed Co. There are two 'Wonder Boys'. One with, and one without VF resistance. Heavy yielder of medium-large (8 oz.), firm fruits, on strong vine with medium foliage cover.

Widely available.

■ *A hybrid with good performance reports.*

	Vineripe	VFN	Hy	80 days	Ind.

Origin: Petoseed Co. One of the better quality hybrid tomatoes. Heavy yields of smooth, medium-firm, large (9 oz.) fruits. Vine is vigorous, with excellent leaf cover.

Available: (21), (26).

■ *Rugged hybrid with triple disease resistance.*

	He Man	VF	Hy	80 days	Ind.

Origin: Goldsmith Seeds, Inc. High yields of medium-firm, medium-sized (6 oz.), uniform, red fruit. Vigorous vines. High yields. Tolerant of verticillium and fusarium wilts. Stake or cage.

A new introduction.

Available: (18), (21), (29), (41).

	Homestead 24	F	S	82 days	Det.

Origin: Asgrow Seed Co. Consistent performer throughout the South. Blossoms set well under a wide variety of conditions, especially high temperatures. Firm, smooth, and meaty, medium-sized (8 oz.) fruits on medium-large vine with dense cover. Resistance to catfacing.

Widely available.

■ *Southern favorite. Recommended by Agricultural Extension Services in AR, FL, GA, OK, VA.*

	Brimmer		S	85 days	Ind.

Origin: Richmond, VA. Large, well-shaped, purplish-pink, 'Beefsteak'-type fruit (to over 1 lb.) is meaty, with few seeds and no core, smoother than the 'Beefsteak' and less likely to catface. Good slicing tomato. Grow in a cage or stake. An old favorite in the Atlanta area.

Available: (15), (22), (32).

	Indian River	F	S	85 days	Ind.

Origin: Florida AES. Scarlet-red, medium-sized (6 oz.) fruit. Thick-walled, firm flesh. Small blossom-end scars. Vigorous vine yields well under warm, humid conditions. Excellent disease resistance.

Plants available at nurseries.

■ *A Florida special. Disease resistance to fusarium wilt, leafmold, and gray leaf spot.*

	Bragger		Hy	87 days	Ind.

Origin: Goldsmith Seeds, Inc. Large (13 oz. and larger) red, meaty fruits. Reported resistance to cracking. Strong-growing vine with large leaves. Recommended pruning to one stem and staking or caging.

A new introduction.
Available: (21), (41).

	Manalucie	F	S	87 days	Ind.

Origin: Gulf Coast Experiment Station. Firm and meaty, thick-walled, medium-sized (7 oz.) fruit. Vine is upright, vigorous, with dense cover. Resistance to foliage diseases, cracking, sunscald, and blossom-end rot. Stake, cage or trellis.

Available: (5), (15), (21), (28), (49), (50).

■ *A Southern favorite. Resistance to fusarium wilt, gray leaf spot, and leafmold. Recommended by Agricultural Services in FL, MD, MS, SC.*

Yellow-orange varieties

For years the yellow varieties have been praised for their mild flavor and low acidity. Catalog writers and consumers erroneously concluded that mild flavor meant low acidity. Actually, the yellow tomatoes are just as acid as the red tomatoes. Today home canners are being warned against even mixing them with red tomatoes in any form of processing. There is no justification for this warning. For comparative figures on acidity see page 64.

Here are the statistics on each variety:

'Manalucie'—an old-timer planted throughout the South.

	Golden Delight		S	65 days	Det.

Origin: South Dakota State University. The only early yellow variety.

Available: (12), (33).

	Jubilee		S	80 days	Ind.

Origin: Burpee Seed Co. The old-timer in the group.
Widely available.

	Sunray	F	S	80 days	Ind.

Origin: USDA. Similar to 'Jubilee' but with resistance to fusarium wilt.

Widely available.

	Golden Boy		Hy	75-80 days	Ind.

Origin: Petoseed. The only hybrid of the group.

Widely available.

	Caro Red		S	80 days	

Origin: Purdue University. 10 times the Provitamin A of standard tomatoes.

Available: (4), (9), (12), (16).

White varieties

The death of the myth that the white tomatoes are "acid-free" creates a void that is hard to fill. We may have set the record straight about the acid content of the white tomatoes (see page 64) but we lost a pleasant reason for growing the white varieties. We haven't grown every white variety, but the ones we have grown were tall-growing, indeterminate vines that needed training on a trellis or in a cage.

'Snowball' (78 days). Available: (10), (26), (38).
'White Beauty' (84 days). Available: (4), (20), (21), (39).
'White Wonder' (85 days). Available: (16).

Paste varieties

Home gardeners who make their own catsup, put up whole tomatoes, or can tomato paste, manage to find room for a vine or two of the paste tomatoes. All bear generous crops of small, pear-shaped fruits on strong-growing vines.

	Nova	VF	S	65 days	Det.

Origin: New York A.E.S. Early 'Roma' type. The choice for short-season areas.
Available: (27).

	Napoli	VF	S	63 days	Medium det.

Large yields of pear to plum-shaped fruit, similar to 'Roma VF' with a greater concentration at maturity. Good cover.
Originator *(55).

	Roma	VF	S	75 days	Det.

Origin: Harris Seed Company. Widely recommended. Strong-growing determinate with heavy foliage cover.
Widely available.

	San Marzano		S	80 days	Ind.

Larger, rectangular, straight-sided pear. Drier than others in the group.
Widely available.

	Chico III	F	S	75 days	Det.

Origin: Texas A.E.S. Compact growth, slightly open. Fleshy fruits can be processed for juice. Chico III will set fruit at high tmperatures (76° to 77° F. night and 96° to 97° F. day). Available: (23), (50).

	Royal Chico	VF	S	75 days	Det.

Origin: Petoseed Co. Growth habit better behaved than that of 'Roma.'

Available: (28).

Small-fruited varieties

The small-fruited 'Red Cherry,' 'Large Red Cherry,' 'Yellow Pear,' 'Yellow Plum,' 'Red Plum,' should be considered as a special class. They are rank indeterminate growers capable of producing fruit under the most adverse conditions. They seem to grow through diseases that cut short the life of the large-fruited varieties and set fruit where other varieties fail. Actually, they are not disease-resistant or immune to blossom drop.

The reason they succeed where others fail is the tremendous number of blossoms per plant. If you lose a few, you still have a good yield. The stamina of these small-fruited varieties is much appreciated in areas too hot for normal tomato growing.

We have grown the 'Large Red Cherry' in a hanging basket indoors in a well-lighted window.

Container varieties

Plant breeders have developed a number of varieties that can be grouped as "container" tomatoes. Actually, any variety can be grown in a container, but the following varieties are well adapted to display in containers in any close-up situation. They are listed in order of size of fruit.

	Tiny Tim		S	55 days	Det.

The smallest-sized fruit of the group, is ¾-inch, scarlet red, on a 15-inch vine. Give it a 6-inch pot or hanging basket, or plant two in an 8-inch pot.

Widely available.

	Salad Top		S	60 days	Det.

A miniature compact vine less than 8 inches tall. Produces a good crop of 1-inch fruits. Give it a 6-inch pot or plant two in an 8-inch pot.

Available as transplants at your nursery.

Originator: *(53).

	Small Fry	VFN	Hy	60 days	Det.

Cherry-type, 1-inch fruit in clusters on a 30- to 40-inch vine. Heavy cropper. Give it a 2- to 5-gallon container and train on a trellis. 1970 All-America Selection.

Also a successful garden tomato for production of clusters of small fruit throughout the season. Give it support of a small cage.

Widely available.

	Sugar Lump		S	70 days	Det.

Cherry-sized 1-inch fruit in clusters on a 30-inch vine. Grow it in a hanging basket or on a short trellis. An unusually sweet tomato.

Available: (2), (16), (21), (36), (39).

	Burgess Early Salad		Hy	45 days	Det.

The vine, only 6 to 8 inches tall, spreads to a width of 2 feet. Very large yield of 250 to 300 1½ x 1¼-inch fruit. Grow in an 8-inch pot or hanging basket.

Available: (4).

	Presto Hybrid		HY	60 days	Det.

Heavy yield of half-dollar-size fruits on a small-leaved, 24-inch rather open vine. Use 24-inch stake or minitrellis in 10- to 12-inch pot or tub.

Recommended as an early variety for the garden. The small vine bears a heavy load of fruit. Train or stake.

Available: (13).

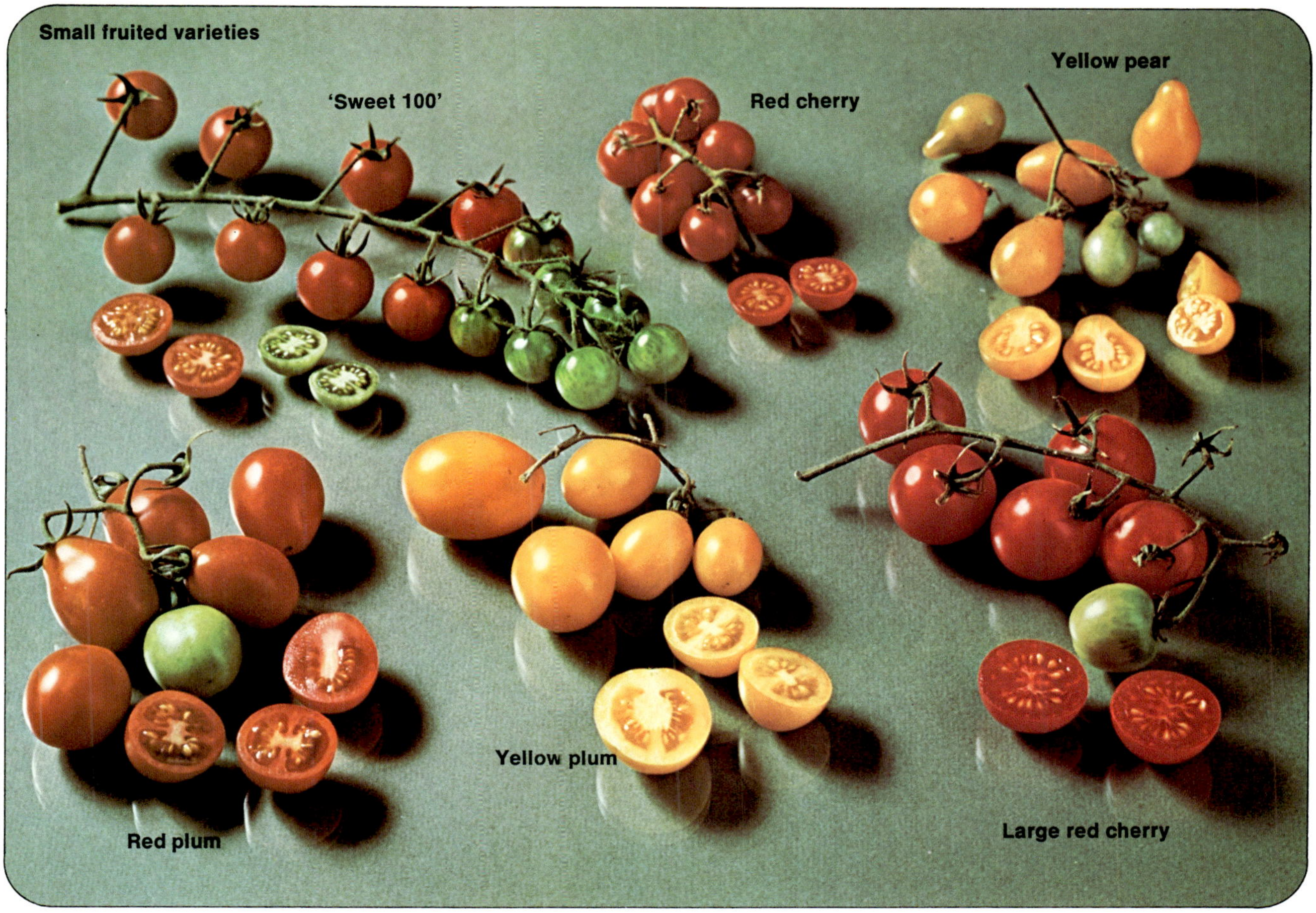
Small fruited varieties
'Sweet 100'
Red cherry
Yellow pear
Red plum
Yellow plum
Large red cherry

Container varieties
'Tiny Tim'
'Tumblin' Tom'
'Pixie'
'Patio'
'Small Fry'
'Sugar Lump'

	Pixie Hybrid		Hy	52 days	Det.

Early ripening, 1¾-inch fruit on 14- to 18-inch plants. Can be grown in an 8-inch pot or hanging basket. Will produce indoors in winter, in a sunny window.

Used also in the garden as the first early variety (52 days). Grow in short cage to prevent sunscald.

Available: (1), (5).

	Toy Boy	VF	Hy	68 days	Det.

Ping-pong-size fruit, on 12- to 14-inch plants. Grow three or four plants in a single 10-inch hanging basket, or pot, or window box. Grow indoors or out with plenty of light.

Available as transplants at your nursery.

Originator: *(56).

	Patio Hybrid	F	Hy	70 days	Det.

Good yield of 2-inch fruits on upright, treelike, 24-inch plant. Will grow in a 12-inch pot but shows off best in a tub or 2-gallon box. Give it support of a short stake when fruit forms.

Widely available.

	Tumblin' Tom		Hy	72 days	Det.

Early and heavy yield of 1½- to 2-inch fruit on 20- to 24-inch vines. Grow in hanging baskets, 2-gallon containers, window boxes.

Available as transplants at your nursery.

Originator: *(53).

	Stakeless	F	S	78 days	Det.

An unusual plant with dense foliage and leaves like those of the potato. Fruits are medium to large (6 to 8 oz.) and of mild flavor. Heavy stemmed, 18-inch, vine needs no staking. We found it most interesting when grown in a box (12 inches wide and 8 inches deep) but gave it the support of short stakes when loaded with fruit. In the garden, in deep soil, it grew 3 feet wide with a heavy yield of fruit.

Available: (6), (27).

Note: Although grouped here as container tomatoes, all are grown in many ways. All are early bearing and excellent choices for climates with insufficient heat to ripen the large, late tomatoes, or in warm climates as an extra early crop of fresh tomatoes preceding the main-season varieties.

Beefsteak varieties

'Beefsteak' is one of the magic words in almost any general discussion of tomato varieties. The word is used to

'Tiny Tim'

'Small Fry,'

Top: 'Pixie' and Bottom: 'Beefsteak.'

describe a type rather than a specific variety—although there is a legitimate variety named 'Beefsteak.'

The picture of the 'Beefsteak' in the minds of the consumer is that of a large tomato in which the flesh is thick and solid with a few small seed cavities.

Where summers are short, the following varieties requiring a warm season of 80 to 90 days are a poor risk. Note the early 'Beefsteaks.'

	**Bush Beefsteak**		**S**	**62 days**	**Compact det.**

Large, (8 oz.) firm fruits. Sets good crops under *most* adverse conditions.

Available: (27), (38), (40), (41), (42).

	**Prime Beefsteak**	**VF**	**Hy**	**70 days**	**Det.**

Origin: Ferry-Morse Seed Co. This VF hybrid earns inclusion of the word 'Beefsteak' in its name because of its meaty character. Smoother than the regular 'Beefsteak.' Large, (8 to 10 oz.), fruit on a vigorous vine.

Originator: *(55).

	**Pink Gourmet**	**F**	**Hy**	**72 days**	**Det.**

Origin: Missouri A.E.S. Good yields of pink, large, firm fruits. The first fruits are rough in shape but smooth out in later yields.

Available: (18), (51).

	**Beefmaster**	**VFN**	**Hy**	**80 days**	**Ind.**

Origin: Petoseed Seed Co. Formerly named 'Beefeater.' Large (to 2 lbs.), fruit on vigorous vine. 'Beefsteak'-type with triple disease resistance.

Widely available.

	**Pink Ponderosa**		**S**	**90 days**	**Ind.**

Old-timer. Fruits very large, (to 2 lbs.), meaty and firm. Large vigorous, open plants with poor cover. Cage for sunscald protection. There's also a 'Red Ponderosa,' similar, except deep scarlet in color.

Widely available.

	**Beefsteak**		**S**	**90 days**	**Ind.**

Formerly called 'Crimson Cushion.' Ribbed fruit, irregular and rough. Vine vigorous with coarse foliage. Best grown in a cage.

Widely available.

Sectional specials

Louisiana

Louisiana State University provides home gardeners with two special varieties with excellent local adaptation: 'Pelican' and 'Creole,' and endorses the variety 'Floralou.'

	**Pelican**	**FN**	**S**	**70 days**	**Ind.**

Origin: Louisiana State University. Of the two, this is the choice for home gardeners who are faced with a nematode problem. Fruit is red, smooth, slightly larger (8 oz.) than 'Creole'and ripens a few days earlier. Sets fruits under higher temperatures than most varieties. Good resistance to foliage diseases. Stake or cage.

A Louisiana special. Available: (22).

	**Creole**	**F**	**S**	**72 days**	**Ind.**

Origin: Louisiana State University. Smooth, medium-sized (7 oz.) red fruit with firm flesh of good texture produced in high yields. Sets fruit in high temperatures. Resistance to foliage diseases and blossom-end rot. Stake or cage.

A Louisiana special. Available: (22).

	**Floralou**	**F**	**Hy**	**74 days**	**Ind.**

Origin: Florida A.E.S. Fruit are smaller than the 'Floradel' but it will out-yield 'Floradel' on the fertile soils of the State. Large yields of medium (5 oz.), unusually uniform, deep red fruit with small seed cavities. Very large vine produces dense foliage cover. Resistance to foliage diseases. Tolerance to cracking and blossom-end rot. Stake, trellis, or cage.

Recommended in Louisiana. Available: (22).

Arkansas

This pink-fruited, Arkansas 'Traveler' has traveled outside the limits of Arkansas.

	**Traveler**	**F**	**S**	**78 days**	**Ind.**

Origin: Arkansas AES. This pink tomato has replaced both the 'Bradley' and 'Gulf State Market 'in commercial plantings in Louisiana. Fruits are of medium size (5 oz.) with good crack resistance. Vines produce good foliage cover. An improved selection has been released by the University of Arkansas with the name 'Traveler 76.'

'Traveler' available: (37).

Texas

To be a success in Texas, a tomato variety should be able to set fruit in high temperatures—90° F. days and 78° F.

nights. Heat resistance is an important factor in the tomato breeding program at Texas A&M University (TAMU).

The varieties 'Porter' and 'Porter Improved' have been Texas home-garden favorites for a long time.

Here are brief descriptions of some of the Texas specials.

	**Plainsman**		**S**	**65 days**	**Det.**

Origin: Texas A.E.S. Medium size, (5 oz.) smooth, firm, bright red fruits borne in clusters that are close together. Very compact plants; can be set 18 inches apart, or grown in small cages. Developed for the high plains of Texas.
Available: (6), (37).

	**Porter**		**S**	**65 days**	**Ind.**

Origin: Porter Seed Co. An old-time favorite. Produces tomatoes about the size of a pullet egg, of good quality and flavor. Fruits have pinkish skin. Relatively large vine with good foliage cover. Porter's catalog describes it as the "old reliable." Produces consistently over a wide range of climatic and soil conditions. Because of its heat resistance, 'Porter' can be started later than other varieties and still produce summer and fall crops.
Available: (30), (37).

	**Porter Improved**		**S**	**65 days**	**Ind.**

Firm, uniform, scarlet-red, medium-sized fruit are smooth and meaty. Vine is vigorous, has good foliage protection, and bears heavy yields until frost.
Available: (37).

	**Gulf State Market**		**S**	**75 days**	**Ind.**

A pink-fruited old-timer. Fairly smooth, small-to-medium sized (4 oz.) fruit is solid, meaty, with few seeds. Flesh is rich, purplish-pink. Fairly tolerant of cracking. Strong-growing vines with good cover bear over a long season.

Recommended in East Texas.
Available: (11), (22).

	**Nematex**	**FN**	**S**	**75 days**	**Det.**

Origin: Texas A.E.S. Nematex is resistant to many disorders of the tomatoes—nematodes, fusarium wilt, early blight, gray leaf spot, and cracking. Heavy yields of medium-sized (6 to 8 oz.), round, smooth, solid fruit. Bush-type vine has sufficient foliage for good fruit protection. Used where nematodes are a problem in Texas.
Available: (37).

	**Supermarket**	**F**	**S**	**75 days**	**Det.**

Origin: Asgrow Seed Co. Smooth, firm, medium-sized (5 to 6 oz.) fruit with very small blossom scar. Resistance to cracking. Fruit has green shoulders which ripen to a rich, uniform red color. Medium-sized vine with fair foliage cover. Very good foliage disease resistance. Train in a short cage.
Available as transplants at garden stores.

Salad tomatoes. Home gardeners have taken the small-fruited varieties developed by Texas A&M primarily for commercial growers, and used them as surefire performers in high-temperature conditions. Gardeners who think that small tomatoes are inferior to the large ones get around their prejudice by thinking of the small tomatoes as "salad tomatoes"—and the larger ones as "slicers."

	**Saladette**	**F**	**S**	**65 days**	**Det.**

Origin: Texas A.E.S. Flavorful fruits about the size and shape of a large plum (average 2 oz.). Thick-walled, meaty, firm, and crack resistant. Small, compact vine, with a 14- to 16-inch spread. Will produce heavy crops under adverse conditions of high heat and humidity.
Available: (30), (37), (41).

	**Chico III**	**F**	**S**	**75 days**	**Det.**

Origin: Texas A.E.S. Small pear-shaped fruits on compact vines. Originally introduced for commercial production, but much used in home gardens as a salad tomato, due to its ability to set fruit in high temperatures. See page 44. "Paste tomatoes."

Are there varieties that adapt to all climates—from humid to arid? Sam Cotner, Vegetable Specialist at Texas A&M, Agricultural Extension Service, answered the question this way:

"If forced to name varieties suited to the many climates of the state, I would first divide the state east and west by drawing a line from Dallas through Waco; East Texas would be east of the line.

"My choice would be:

East Texas		**West Texas**
Better Boy	Homestead 24	Bonus
Big Set	Walter	Walter
Porter	Saladette	Big Set
Spring Giant	Chico III	Saladette
Springset	Gulf States Mkt.	Chico III
Terrific	Supermarket	Spring Giant

"But forgetting this division of the State, to home gardeners who ask, 'What variety should I grow in Texas?' (local address unknown), my list of reliable varieties would be:

Spring Giant	Big Set
Terrific	Bonus
Walter	Homestead 24"

Particulars on all varieties listed are given on preceding pages.

Seed company specials

If you are a reader of seed catalogs you know that almost every seed company has its exclusive tomato variety. These varieties have earned the stamp of approval of the seed company because customer satisfaction results in repeat business.

Some of the seed company specials are:

Improved Wayahead		**S**	**63 days**	**Ind.**

A Jung Seed Co. special.
Fruits are smooth, solid, and of good size for an early variety. Vine bears early and continues to produce through the season.
Available: (16).

Vigor Boy	**VFN**	**Hy**	**63 days**	**Det.**

A Gurney Seed Co. special.
Heavy yields of smooth, meaty, even-ripening fruits, average to medium-sized (6 oz.), and can weigh up to 1 lb.
Available: (12).

Hy-X		**Hy**	**67 days**	**Det.**

A Henry Field Seed Co. special.
Large yields of bright scarlet, medium-sized (4 to 6 oz.) fruit on short, bushy vine. Does especially well in semiarid climates.
Available: (10).

Whopper	**VFN**	**Hy**	**75 days**	**Ind.**

A Park Seed Co. special.
Heavy yields of large (10 oz.), smooth, firm, meaty, fruits. Vine is vigorous and tall-growing, with good foliage cover. Good resistance to cracking and blossom-end rot. Widely adapted.
Available: (21).

Abraham Lincoln		**Hy**	**78 days**	**Ind.**

A R. H. Shumway Seed Co. special.
Origin: Buckbee's Rockford Seed Farms, Illinois.
An old-timer with very large (1 to 2+ lbs.), solid, firm, and meaty fruits. As many as nine in a cluster. Vine is sturdy-growing and continues bearing until frost. For such a large tomato it is remarkably smooth and free from cracks and seams.
Available: (26).

Red Chief	**VFN**	**Hy**	**80 days**	**Ind.**

A Hastings Seed Co. special.
Excellent wilt and nematode resistance. Large (9 oz.), very smooth fruit. Vine is vigorous and productive over a long season. Stake, trellis, or cage.
Available: (15).

Varieties differ in leaf shapes

Left: Hand watering can become a chore when outside temperatures soar. Center: The greenhouse gardener's goal: your own fresh vegetables the year 'round. Right: Heavy tomato yields are possible in April from late winter plantings.

Greenhouse growing

Home-garden greenhouses are used in many ways. The most common use is for the early start of flowers and vegetables from seed. The greenhouse is a necessity to the hobbyist specializing in the exotic—orchids, ferns, and foliage plants. Using the greenhouse to grow vegetables in the off-season months is on the increase.

The home gardener with an interest in growing greenhouse tomatoes commercially should realize that growing tomatoes in a greenhouse for home consumption and growing for profit are entirely different operations.

If you are growing commercially you grow what you can sell, and what you can sell will be governed by local buying habits and ideas about what a greenhouse tomato should look like.

Dr. Victor Lambeth of Missouri State University—the father of the 'Tuckcross' varieties of greenhouse tomatoes—talked about the strange pattern of regional acceptance of size and color:

"Let me give you an observation on greenhouse tomatoes as per size of fruit. The New England States want a small fruit, similar to what they grow in England. In Europe the preferred greenhouse tomatoes will not weigh over 3 or 4 ounces. But you can't get a Midwesterner to look at them.

"And there is a strong bias with respect to the skin color. There are certain areas that will accept a pink and not a red, and there are others that will accept red and not the pink. It's just like white eggs and brown eggs.

"In Canada some greenhouse growers are breaking away from the small tomato preference and breaking into the larger fruit market.

"As you go into the Deep South, winter low-light conditions improve, and you can grow the field varieties in greenhouse structures, particularly the Florida varieties like 'Tropic,' and 'Manapal,' and do quite well with them.

"In the West the greenhouse varieties most commonly available at supermarkets are the hydroponically grown, large-fruited 'Tropic' and 'Manapal.' They are usually identified by individual labels, as well as a difference in price."

The United Fresh Fruit and Vegetable Association sheds some light on why some stem is left on the fruit of the greenhouse tomato when harvested.

"On the market, greenhouse tomatoes are judged somewhat by the freshness and greenness of their sepals, as this is an index to the length of time the fruits have been removed from the vines. Fresh sepals on full-ripe fruits indicate they were recently harvested when practically ripe. Tomatoes with stems attached lose moisture more slowly and will keep fresh for a longer time than those with stems removed."

For the amateur

The increased interest in greenhouse growing has given the home gardener a wide choice of commercially available greenhouse structures. It is generally agreed that the beginning greenhouse gardener should start with one of the smaller, commercially available greenhouses, practice with it for a season or two, and then branch out to a larger model, or if desired, build a custom-designed unit.

The type of structure and the heating and cooling equipment required will be determined by your climate. Summer heat and winter cold are the two primary considerations.

Obviously, those in mild-winter areas with a high percentage of sunlight find off-season vegetable growing far more practical than greenhouse growers in severely cold winter areas.

To cut back on heating costs in harsh winter climates, greenhouses should be double glazed, or designed in such a way so that a second layer of plastic film can be easily installed.

In areas where summers are hot, temperatures inside the greenhouse may reach levels unfavorable to blossom set. If high humidity is not a problem an individual cooling unit, such as a "swamp cooler," or a "wet wall" unit will bring temperatures down to an acceptable range.

Left: Greenhouse plants trained to a single stem supported by plastic twine. Center: Bottom end of twine tied to base of plant; top end clipped to overhead wire. Right: One of the favorites of the greenhouse grower is the variety 'Floradel.'

The wet wall system, which has proven itself effective, especially in larger greenhouses, is made up of a series of wood fiber pads installed vertically on one side of the greenhouse and large exhaust fans on the other side. The exhaust fans draw air through the pads, which are kept wet by continually circulating water. The moist air coming through the pads is cooled by evaporation of the water.

If summer temperatures *and* high humidity are a problem, swamp coolers and wet wall systems will only add to the humidity problem. A refrigeration unit of some kind is recommended in these areas.

In areas where summer temperatures are not extreme, but heat build-up is still a concern, temperatures inside the greenhouse can be lowered by temporarily (usually from May to October) removing the sides of the greenhouse, and replacing them with a wire screen, such as chicken wire, to keep animals out.

Growing methods

Over the years several different growing methods have been practiced by commercial greenhouse tomato growers. Home greenhouse gardeners can benefit from their experiences when choosing the method best for their own situation. There are four more or less accepted methods.

Bed culture. Tomatoes are planted directly in the soil on which the greenhouse structure is built. The enrichment of the soil with organic matter and fertilizers is much the same as with gardening outdoors. The big drawback for commercial growers, and even more of a handicap for home gardeners, is the necessity for yearly soil sterilization to control soil-borne diseases, particularly harmful to tomatoes.

Trough, ring, and hydroponic growing of greenhouse tomatoes are newer growing methods introduced primarily to eliminate or reduce the need of sterilizing entire greenhouse ground beds. The principles involved are basically the same—the using of a sterile growing medium, and confining the roots to a relatively small area.

Trough growing. Tomatoes are grown in long, narrow, plastic lined beds, filled with a lightweight soil mix. In order to be effective, the plastic liner must be impermeable to roots. The troughs can be constructed out of concrete blocks, 1 x 6- or 2 x 6-inch lumber or similar building materials. Troughs should be 5 to 6 inches deep, and about 24 inches wide to accommodate two rows of tomatoes. Drain holes, 1 inch in diameter, should be cut on each side of the plastic liner, 1 to 2 inches from the bottom, about ten feet apart.

Ring culture. The tomato plant is set into a round (8- to 10-inch-diameter) ring, or sleeve, of plastic film or tar paper. The rings have no top or bottom and are filled with a sterile growing medium and placed on a troughlike bed of water-absorbing aggregate, 4 to 6 inches deep. Ring culture was originally promoted because it allows the tomato plant to form two separate root systems. It is claimed that by planting in an artificial mix (in the ring) and allowing the roots free access into the water-charged underlayer, the combined advantages of soil culture and hydroculture may be realized.

Hydroponic culture is a nonsoil system of growing tomato plants. The tomatoes are grown in specially designed tanks on beds filled with aggregate or similar material. Irrigation is taken care of by an automated battery of pumps, with regularly timed nutrient applications.

Timing

In commercial production the late spring single crop is by far the easiest greenhouse tomato crop to produce and is highly recommended for the amateur greenhouse gardener. Seeds are planted in late winter, and the plants set permanently in the beds in early spring. Flowering, fruit setting, and fruit growth occur during the increasingly longer and sunnier days of spring and early summer. Harvest begins much in advance of field-grown varieties, and extends into the season until the field varieties are commonly available. Excellent crops of high-quality fruit are possible with plants producing from eight to twelve clusters with one to two pounds of fruit per cluster.

'Vendor' (left) and 'Tropic' (right) are recommended for both greenhouse and home garden plantings.

Pruning and training

Plants, regardless of the growing method, are usually trained to a single stem. All sideshoots should be removed at least twice weekly. Plants are supported with plastic twine, one end of which is tied with a small nonslip loop to the base of the plant. The other end is attached to a wire supported six to eight feet above the plant row. As the plant grows, it is twisted around the twine in one or two easy spirals for each cluster of fruit.

Varieties

Here are the varieties for greenhouse growing that are available through mail-order catalogs. In areas of high percentage of winter sunshine many of the productive field varieties are being grown successfully in a greenhouse.

	Michigan-Ohio	**F**	**Hy**	**70-75 days**	**Ind.**

Origin: Michigan A.E.S. "The most widely used and successful large red tomato for growing under glass or in plastic houses." Especially developed for low-light greenhouse conditions. High yields of medium-large (8 oz.) fruit with thick walls on strong-growing, dark green, vigorous vines. Its chief limitation is lack of resistance to leaf mold, and in some areas it tends to blotch. Widely available.

	Ohio-Indiana	**F**	**Hy**	**74 days**	**Ind.**

Origin: Ohio A.E.S. Medium-sized pink fruit. Vine is very productive. Resistance to leaf mold. Customers in southern Ohio accept a red tomato; while those in the northern part of the state prefer a pink variety like this. Available: (21), (36), (40).

	Tuckcross 520	**F**	**Hy**	**74-82 days**	**Ind.**

Origin: Missouri A.E.S. One of the latest of the 'Tuckcross' series. Slightly earlier than 'Michigan-Ohio Hybrid.' In addition to resistance to fusarium wilt, it has greatly improved resistance to races of leaf mold, has larger red fruit (6 to 8 oz.) and does not show the tendency to "over-flower" and "over-set," typical of the older 'Tuckcross Hybrids.' Used both as a fall and spring crop. Will set fruit under adverse conditions. Available: (21), (28), (37), (40).

	Tuckcross 533	**F**	**Hy**	**72-82 days**	**Ind.**

Origin: Missouri A.E.S. Very similar to 'Tuckcross 520.' Tends to set fewer fruit, but larger ones. Also has resistance to leaf mold. Available: (27), (37), (40), (49).

	Vendor		**S**	**63-65 days**	**Ind.**

Origin: Vineland Hort. Res. Inst., Ontario, Canada. Heavy yields of bright red, very firm, medium-sized (6 to 8 oz.) fruit on sturdy plant. Ripens uniformly. Plant is slightly shorter than most greenhouse types, and fruit clusters closer together than most varieties. Easy to prune and pollinate. Small immature fruit sometimes appear egg-shaped, but fill out at maturity to a deep, globe-shaped fruit. Resistance to leaf mold and cracking. A report reads: "Shows good resistance to tobacco mosaic virus in Canada, but not in Ohio." Highly recommended for home garden growing. Available: (27), (36), (37), (40). (36), (37), (40).

	Tropic	**VFN**	**S**	**81 days**	**Ind.**

Developed by the University of Florida. This very strong newcomer is one of the most popular greenhouse varieties. Good yields of very firm, large (8 to 9 oz.) fruit, on a vigorous, strong-growing vine with good foliage cover. Sets large fruit high on the vine to the end of the growing season.

Multiple resistance to foliage diseases includes: gray leaf spot, greywall, leafmold, early blight, and tolerance to races of tobacco mosaic. Also, resistance to blossom-end rot. Grow in a cage, on a trellis, or stake. Best as a pink harvest. Also highly recommended for home garden growing. Available: (2), (5), (27), (28), (37).

	Manapal	F	S	75 days	Ind.

Origin: Florida A.E.S. Very widely adapted. Smooth, firm, medium-sized (6 oz.) thick-walled fruit on a large indeterminate plant with heavy foliage cover. Has resistance to fusarium wilt, gray leaf spot, and leaf mold. Grows best in a cage, on a trellis, or staked. Also highly recommended for home garden growing. Available: (28), (30), (37).

	Floradel	F	S	78 days	Ind.

Origin: Florida A.E.S. Fruit is medium-sized (6 oz.), with thick walls, smooth and firm. Vine has large leaves, and provides good cover. The fruit sets well at low temperatures. Resistant to gray leaf spot, and leaf mold. Grow on a stake, cage, or trellis. Also, highly recommended for home garden growing. Available: (22), (23), (30), (37).

Other new varieties are being introduced. Ohio A.E.S. has introduced tobacco mosaic-resistant varieties: 'MR 12' and 'MR 13.' These advanced disease-resistant varieties are not available through the seed companies on our list.

Inform yourself

Check the advantages and disadvantages of the various designs of greenhouses in operation in your area. How are the small hydroponic units performing? What about increases in the heating bill? Check with your County Extension Agent about the problem of growing tomatoes in the home-garden greenhouse. Get the greenhouse publications available at the office of your County Extension Agent. Write for publications in the following list:

A Simple Rigid Frame Greenhouse for Home Gardeners. Circular 880, 2/64. Agricultural Publications Office, University of Illinois, Urbana, IL 61801. (15¢)

Building Hobby Greenhouses. Bulletin #357, 3/73. Superintendent of Documents, U.S. Government Printing Office, Washington, DC 20402. (35¢)

Building Hobby Greenhouses. #AAB357. Agricultural Bulletin Building, University of Wisconsin, 1353 Observatory Drive, Madison, WI 53706. (15¢)

Plastic Greenhouse Manual—Planning, Construction, and Operation. by Raymond Sheldrake, Jr., and Robert M. Sayles, 1974. Cornell University, c/o Dr. Raymond Sheldrake, Vegetable Crops Dept., Ithaca, NY 14853. ($2)

A Gothic Greenhouse for Town and Country Homes. Publication 487, 3/72. Extension Division, Virginia Polytechnic Institute and State University, Blacksburg, VA 24061.

Plastic Covered Greenhouse. USDA Misc. Publications #1202, 7/71. Superintendent of Documents, U.S. Government Printing Office, Washington, DC 20402. (35¢)

Design and Operation of Greenhouse Cooling System. Bulletin #AENG1, 7/68. Extension Agricultural Engineer, Texas A&M University, College Station, TX 77843.

Small Greenhouses Provide Year-Round Pleasure. Yearbook Separate No. 3805, 1972. USDA-ARS, Building 046-A Arcw, Beltsville, MD 20705.

Home Greenhouses for Year-Round Gardening Pleasure. Circular 879. Cooperative Extension Service, College of Agriculture. University of Illinois, Urbana, IL 61801. (20¢)

Growing Greenhouse Tomatoes in Ohio. #SB-19. Cooperative Extension Service, Ohio State University, 2120 Fyffe Road, Columbus, OH 43210. ($2)

Production of Greenhouse Tomatoes in Ring Culture or in Trough Culture. Mimeo No. 149, 3/69. Department of Vegetable Crops, New York State College of Agriculture, Cornell University, Ithaca, NY 14850.

Nutriculture Systems, growing plants without soil. Experiment Station Bulletin #44, 3/74. Department of Horticulture, Agricultural Experiment Station, Purdue University, West Lafayette, IN 49707.

Hydroponics as a Hobby. Circular 844, 3/72. Cooperative Extension Service, Univ. of Illinois, Urbana, IL 61801. (15¢)

Hobby greenhouses & other gardening structures. 60-page booklet. Order from: NRAES, (Northeast Regional Agricultural Engineering Service), Riley-Robb Hall, Cornell University, Ithaca, NY 14853. ($2)

A series of greenhouse publications. The University of Kentucky has made available a packet of valuable and numerous publications covering every aspect of greenhouse building and environment control. Write to: Bulletin Room, Ag. Exp. Station Bldg., University of Kentucky, Lexington, KY 40506.

First copy is free, multiple or quantity copies are priced according to list below:

AEN—6 "Preservative Treatment of Greenhouse Wood," 5¢ each.
AEN—7 "Polytube Heating-Ventilation Systems and Equipment," 10¢ each.
AEN—8 "Estimating Greenhouse Heating Requirements and Fuel Costs," 5¢ each.
AEN—9 "Estimating Greenhouse Ventilation Requirements," 5¢ each.
AEN—10 "Greenhouse Coverings," 10¢ each.
AEN—12 "Greenhouse Structures," 10¢ each.
AEN—13 "Greenhouse Benches," 10¢ each.
AEN—15 "Rigid-Frame Greenhouse Construction," 7¢ each.
AEN—18 "Air Circulation in Greenhouses," 10¢ each.
AEN—19 "Greenhouse Humidity Control," 15¢ each.
AEN—32 "Greenhouse Location and Orientation," 15¢ each.
MISC. 395 "A Greenhouse for Teaching Plant Science and Production," 15¢ each.
AEN—28 "Cooling Greenhouses," 15¢ each.
AEN—30 "Greenhouse Heating Systems," 15¢ each.
AEN—31 "Greenhouse Heating Systems," 15¢ each.
AEN—14 "Painting Greenhouses and Equipment," 10¢ each.
AEN—4 "List of Plans for Greenhouse & Horticultural Facilities," 7¢ each.

Greenhouse information packet. Pennsylvania State University makes available a packet of publications on greenhouse building and management. Write to: Agricultural Engineering Extension, 204 Agricultural Engineering Bldg., University Park, PA 16802. ($1)

For the "soil mix" test, we made sure that no vine had any special planting advantage. Transplants of the same size were planted the same depth.

These photos, taken at 35 days (above) and 98 days (below) show the superior vigor of the synthetic soil mix grown tomatoes. #1 through #4 are growing in top soil; #5 and #6 in a synthetic mix.

Experiments

With the deck, balcony, and patio gardeners in mind, we directed our experiments to answer some of the questions frequently asked about growing tomatoes in containers:

✔Do you need a special soil mix?
✔What size container is best?
✔What about training the vine?

Container soil-mix test

We tested soil mixes in our Los Angeles test garden. The objective of the test was to compare the locally available top soil to quality controlled synthetic soil mixes.

The tomato variety chosen for the test was 'Better Boy.' Transplants were purchased at a local nursery and planted out on April 20 in a series of containers 24 inches square by 12 inches deep. Containers 1 through 4 (see photograph) were filled with topsoil. Containers 5 and 6 were two of those filled with the synthetic soil mix.

The fertilizing, mulching and watering program was the same for all plants throughout the experiment.

Results. The tomatoes grown in topsoil showed water stress, but fruited and ripened earlier. The plants grown in synthetic soil produced twice as much fruit per plant.

	Top soil	Synthetic soil mix
Days to flower	20	20
Days to set fruit	42	49
Days to ripe fruit	84	94
Lbs. of ripe fruit @ 98 days	4.5	9.0

How much soil?

We have long debated the minimum volume of soil required to grow a quality crop of tomatoes in containers. In this experiment we tried three different amounts of our standard soil mix—4 cubic feet, 2 cubic feet, and ½ cubic foot.

We planted 'Better Boy' tomatoes in three different sized containers in the same soil mix on the same day.

All were watered and fertilized in the same way. Only the volume of soil varied.

45 days after planting. We compared the plants 45 days after planting. All the plants were thriving—green leaves, flowers, and even some fruit set. The major difference was in the height of the plants.

45 days			
Soil volume	4 cu. ft.	2 cu. ft.	½ cu. ft.
Height	30″	27″	21″

Clearly the tomato plants in ½ cubic foot containers were lagging in height.

90 days after planting. After another 6 weeks we measured the plants. Again, all the plants showed good vigor, strong new growth, and continued fruit set. This time the difference among the tree soil volumes showed in both height and the load of ripening fruit.

90 days			
Soil volume	4 cu. ft.	2 cu. ft.	½ cu. ft.
Height	66″	56″	46″
Fruit	58	48	12

The chart clearly shows that the larger the volume of soil, the larger the tomato harvest. But, if you're talking about maximum production from the minimum amount of soil, bear in mind that a 4-cubic-foot container holds enough soil to fill eight 5-gallon containers. If you have the deck space for eight 5-gallon containers, the total yield from those eight plants will be almost double that of the plants in the 4-cubic-foot box.

Trellis training

We wanted to see how much production we could get from a container filled with ½ cubic foot of soil mix, the equivalent of a 5-gallon can. The *only* difference between this experiment and the last one is the training method. The varieties 'Big Set,' 'Vineripe,' 'Springset,' and 'Tropic' were all trained on either a wire or lath trellis. Pruning was kept to a bare minimum—all shoots being tied to the trellis.

In the case of 'Big Set' (see photographs on the next page), simple boxes were built, 12 inches square, by 8 inches deep. The other varieties were planted in 5-gallon cans.

45 days after planting. We looked at our plants 6 weeks after planting and noted good growth, flowers, and fruit set. The first two flower clusters had set fruit, and the clusters were full.

90 days after planting. The plants were vigorously growing. Flowers were blooming on the new growth, young fruits were setting, and the lower leaves were still a rich green color. The best performer of the four varieties had 31 ripening fruit, the lowest had 25. The average was 29 fruit per plant. We have already picked three dozen ripe fruit, and there's more coming.

# Fruit at 90 days	
'Big Set'	31
'Springset'	32
'Vineripe'	25
'Tropic'	29
Average	29

The variety 'Better Boy' grows in a large container holding 4 cubic feet of soil; box with 2 cubic feet of soil; in a can holding ½ cubic foot. The difference in growth is apparent, but the winner of yield of fruit is the can with only ½ cubic foot of soil. See text.

Screen of lath covered with cheesecloth on three sides of a 2 cubic foot box supports and protects vines of 'Better Boy.'

The variety 'Big Set' grows in boxes holding ½ cubic foot of soil—the same amount as the 5-gallon nursery can. Each vine is trained fan-shaped on a wood trellis. Yield of fruit was spectacular. Was the mulch of the automator responsible? It played a part, but vines in 5-gallon cans without mulch performed equally well. See photograph below.

Vines of the varieties 'Vineripe,' 'Tropic,' and 'Springset' are trained on lath trellis on deck. All are growing in 5-gallon nursery cans (½ cubic foot of soil). Cans are hidden from view with a short lath screen. Fruit yields more than satisfactory.

Other training notes

A tomato's performance in a 5-gallon container is strongly influenced by the training method. In the 'Better Boy' experiment, where the 5-gallon container is compared to the larger sized containers, the tomato was pruned and trained on a tomato tower. (See top photo, page 55). This method produced larger and earlier fruit than the plants in the larger containers. However, in all tests of production in 5-gallon containers, the number of fruit is greater when the vine is trained in a fan shape, on a trellis of wire or wood.

What advantage in deep planting?

The general rule in transplanting is "never plant deeper than it grew in the nursery." The tomato plant has the ability to develop roots along the stem, wherever the stem comes in contact with moist soil. For that reason set any tomato transplant deep in the soil—up to its first leaves.

To find the difference in growth of the deep planted tomato, and the one planted normally, we planted the same variety both ways. The results are shown in the photographs below.

Are the experiments valid?

Can the experiments be repeated in another garden situation with comparable results? Here are the factors we controlled in the last two experiments, and the climate we live with.

Special soil mix. The soil mix was made up of 45% sphagnum peat moss, 45% perlite, and 10% sand. To one cubic yard of this mix these fertilizers and amendments were added:

- 8 lbs. of 5-10-10 fertilizer
- 10 lbs. of agricultural lime

Any of the quality controlled commercial mixes such as Jiffy Mix, Redi-Earth, or Super-Soil will give you equal, if not better results.

Temperature and sunlight. The factors of sunlight and daily temperature will, of course, vary from garden to garden. We would like to say that every tomato plant used in any of our experiments received the same amount of sunlight. They didn't.

Some get the morning sun, some the afternoon sun. Some vines that received 7 hours of sunlight in June are in shade most of the day in late September.

As with all gardens everywhere, as the garden grows older the sunny spots become fewer. The shade tree spreads wider and wider. Your neighbor's hedge blocks the late afternoon sun.

Are the differences in the amount and quality of the sunlight received by the vines a significant factor in our measurements of performance?

The vines in the 5-gallon containers with a trellis receiving 7 hours sun have given us more ripe fruit than the vines in the 5-gallon containers in the shady locations, but both are heavy with fruit.

In sun or partial sun the vines produce a satisfactory number of tomatoes.

The temperatures were generally in the mid 70's and low 80's. As always there were wide variations; this year an early heat wave hit when our container tomatoes were about 18 inches high.

For 6 days the temperature exceeded 97°; on 4 of those days it reached 102°. Until this time, the tomatoes had been growing well in the early spring weather, but how would they stand the sudden 20° to 30° jump in temperature?

We poured the water to them—filling the container until water ran out the drainage holes. This increased watering got us through the heat wave, but it was the porous soil mix that allowed us to water so frequently, without fear of waterlogging the roots of the plants.

The tomatoes withstood the heat without wilt. That week's flowers failed to set fruit, but when the temperatures returned to the mid 70's the fruit set without a hitch.

Deep planted transplant forms roots both from the rootball and all along the buried stem.

Vine of the deep planted transplant photographed 30 days after planting.

Photographed on the same day is the vine of the transplant planted at the depth it grew in the nursery.

Chromosome map of tomato

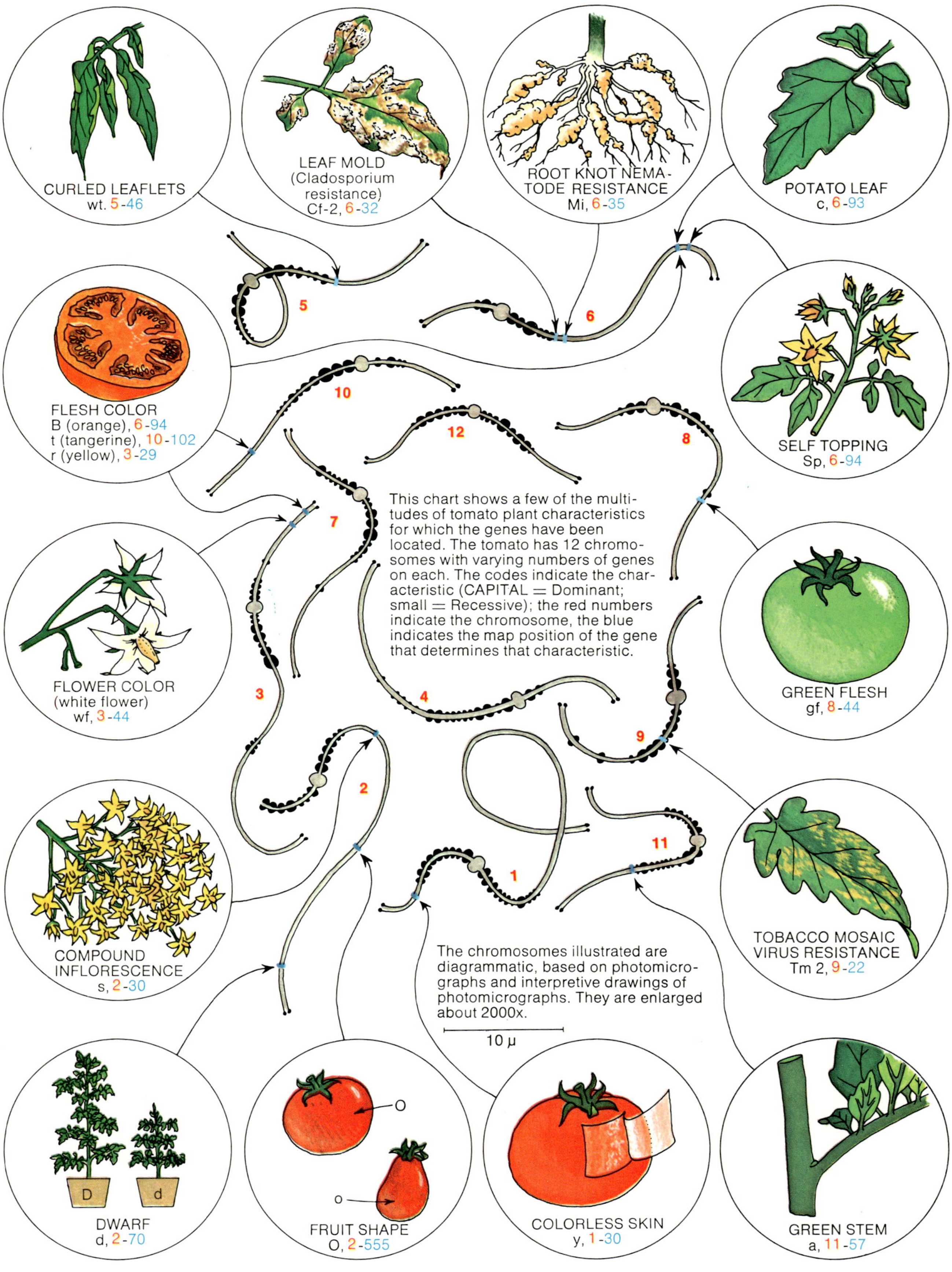

This chart shows a few of the multitudes of tomato plant characteristics for which the genes have been located. The tomato has 12 chromosomes with varying numbers of genes on each. The codes indicate the characteristic (CAPITAL = Dominant; small = Recessive); the red numbers indicate the chromosome, the blue indicates the map position of the gene that determines that characteristic.

The chromosomes illustrated are diagrammatic, based on photomicrographs and interpretive drawings of photomicrographs. They are enlarged about 2000x.

Trouble shooting

In this short chapter we deal with the major troubles your tomatoes may encounter. We'll show you how to prevent problems, and how to handle trouble when you can't avoid it.

To the home gardener in trouble: The tomato breeder is coming to your rescue.

The chromosome map on the adjacent page shows the inner workings of the tomato. You don't have to understand the map to know that it spells help to gardeners who have had trouble growing tomatoes, especially trouble with diseases.

The chromosome map illustrates the twelve chromosomes of a tomato plant. Plant breeders and researchers specializing in tomatoes have isolated many of the particular characteristics of the tomato, and discovered the location of the gene responsible for that characteristic on individual chromosomes.

Tomatoes with increased disease resistance and wider adaptation are on their way as more characteristics are mapped, or located on the tomato's chromosomes. At this time it is possible to breed into a single tomato variety resistance to, or tolerance of, fifteen different diseases and disorders.

Not yet on the genetic map, but a very important characteristic in breeding lines is resistance to Verticillium wilt and Fusaruim wilt—two of the most troublesome diseases of tomatoes.

"V," "F," and "N"

The best way to avoid problems due to Verticullium wilt, Fusarium wilt, and nematodes, is to plant disease-resistant varieties.

If you have Verticillium wilt you can still get tomatoes, but Fusarium wilt is more serious and wipes out the entire crop.

Fusarium and Verticillium wilts may occur in the same soil, and at times are hard to distinguish from one another. Typical cases of Fusarium wilt are distinguished by the yellowing of leaves on single branches, frequent wilting and death of the plant before the end of the season. The progress of Verticillium wilt is favored by temperatures of 70° to 75° F. and retarded by the higher temperatures that are most favorable to Fusarium wilt.

The resistance to disease is a part of the name of many varieties. For example: 'Wonder Boy VFN,' 'Jet Star VF,' and 'Burpee VF.' "V," "F," and "N" are the initials to look for when choosing varieties, if you've had trouble with these diseases or nematodes in the past.

Verticillium wilt

The first symptom is yellowing of the older leaves, accompanied by a slight wilting of the tips of the shoots during the day. The older yellowed leaves gradually wither and drop, and eventually the crown of the plant is defoliated. The leaves higher up the stem become dull looking, and the leaflets tend to curl upward.

All branches are uniformly affected and have a tendency to be less erect than those of healthy plants. The plants usually live through the season but are somewhat stunted, and the fruits are small. In late stages of the disease, only the leaves near the tips of the branches remain alive. The loss of the lower leaves and the stunting of the later growth expose the fruit to the sun, and much of the crop is often lost because of sunscald.

Fusarium wilt

According to USDA studies, Fusarium wilt is one of the most prevalent and damaging diseases of tomatoes in many of the tomato-growing regions of the United States. The fungus is soil-borne, found in both field and greenhouse soils. The organism generally does not cause serious losses unless soil and air temperature are rather high (80° to 90° F.) during much of the season. Where such conditions prevail, plants of susceptible varieties either are killed or severely damaged when grown in infested soil.

In seedling plants Fusarium wilt causes drooping and downward curvature of the oldest leaves, usually followed by wilting and death of the plant. Older plants in the field are infected at all stages of growth, but the disease generally becomes most evident when the plant is beginning to mature its fruit. The earliest symptom is yellowing of the lower leaves. Often the leaves on only one side of the stem turn yellow at first; also the leaflets on the side of the petiole may be affected before the others. The yellowed leaves gradually wilt and die and, as the disease progresses, yellowing and wilting continue up the stem until the foliage is killed and the stem dies.

Frequently a single shoot is killed before the rest of the plant shows much injury. The stem of a wilted plant shows no soft decay, but if it is cut lengthwise, the woody part next to the green outer cortex shows a dark brown discoloration of the water-conducting tissues.

Nematodes

Tiny parasitic eelworms called nematodes cause severe damage to tomato plants. Their presence should be suspected when poor stands, stunted plants, wilting of some plants more than others, and plant death are noticed. Plant injury can usually be found by examining the roots for swollen, knotty galls (see illustration), or brown, sheared-off areas.

Controlling nematodes is practical and inexpensive using the in-the-row fumigation method. Open a trench six-inches deep where tomatoes are to be planted. Dribble in the fumigant at the manufacturer's recommended rate. Fill the trench immediately, and wet the surface with water to seal the fumigant in the soil. Allow sufficient time before planting for proper aeration. Read and follow all label directions.

Among the products the University of Florida recommends for fumigating soil prior to planting, are: Dowfume W-85, Soilbrom 85, Vidden D, Fumazine 86E, Nemagon 50%, Nemagon 12EC, Oxy BBC 12E, Vapam, VPM, and Vorlex.

The University of Florida also states: "Where nematode injury is suspected after tomato plants are up and growing, Nemagon (granules or liquid) may be applied to the root zone. However, such post-transplanting applications may cause injury to the plants and should be made only as a last resort."

Physiological disorders

Besides the triple threat of "V," "F," and "N," there are a lot more problems and troubles you might run into that can plague the tomato plant and/or fruit.

Stress

Stress and stressed are useful words to the tomato worker's vocabulary.

When a plant is *stressed* for water and then heavily watered, you can expect blossom-end rot.

High temperatures will put the tomato plant under *stress* not only in water deficiency but in the normal coloring process.

It all goes back to the fundamentals: give the plant what it needs, a constant supply of moisture, and a constant supply of nutrients.

Blossom-end rot

Symptoms of this disease appear as a leathery scar or rot on the blossom-end of fruits. Blossom-end rot can occur at any stage of development. It is usually caused by sudden changes in soil moisture. It is most serious when plants growing rapidly with high soil moisture are hit by a hot (temperatures above 90° F.) dry spell. Lack of calcium in the plant is another cause of blossom-end rot. (See page 20 for ways to add calcium to soil.)

Mulch plants with black plastic or organic material to reduce fluctuations in soil moisture and temperature. Do not plant in poorly drained soil.

Staked and heavily pruned tomatoes seem more susceptible to blossom-end rot than plants trained without pruning.

Blossom drop

This frustrating habit of the tomato plant is discussed in detail on page 11. Throughout the presentation of varieties (pages 34-49), we note the varieties with the ability to set fruit under adverse conditions.

Tomato gardeners have worked out many ways to increase blossom set. These have been successful:

✔In early spring, increase night temperatures by covering the young plants with plastic tents or similar coverings. (See page 22.)

✔Protection of the plants from strong winds, using windbreaks. (See page 23.)

✔Daily vibration of the individual flower clusters to increase fertilization.

Catfacing

Cool and cloudy weather occuring at the time of bloom may cause the blossom to stick to the small fruits, resulting in the malformation of the fruits as they develop. The blossom ends of fruit are puckered, with scars between the lumps. Cavities may deeply penetrate the fruit.

Catfacing occurs most frequently on early harvest, usually on the large-fruited tomato varieties.

The best control is to plant varieties resistant to catfacing. See pages 34-49.

Fruit cracks

There are two distinct types of fruit cracks—radial and concentric (see illustration). Radial cracking is most common and results in most fruit damage. It occurs more often during rainy periods when the temperature is relatively high (above 90° F.), especially when rains follow a long dry period—conditions which promote extremely rapid growth. Radial cracking is most severe on ripening and full-ripe fruits. Tomatoes exposed to the sun develop more cracks than those well-covered with foliage.

The use of crack-resistant varieties and maintaining a uniform water supply through the use of irrigation or mulches, and keeping a good foliage cover will help reduce cracking. See pages 34-49, for crack-resistant varieties.

Leaf roll

The leaf roll we are talking about here is not the leaf roll caused by disease. This type of leaf roll is caused by prolonged rains causing too much moisture in the soil, or by severe pruning. Some tomato varieties are more inclined to roll than others.

Lower leaves are the first to roll. The rolling continues up the plant. Leaflets may cup and overlap. Leaves feel firm and leathery.

The best control is to maintain a uniform soil moisture—by using a plastic mulch and irrigating regularly.

Misshapen fruit

If you grow many tomatoes you know the meaning of abnormal or misshapen fruits. There are bulges and whatnots on varieties that should be round and smooth.

Young plants exposed to temperatures of 55° F. or lower will usually produce rough or misshapen fruit on the first cluster. These low temperatures interfere with the growth of pollen tubes and normal fertilization of the ovary.

The more common abnormalities are pointed fruits with an elongated blossom end, or puffy fruit in which air spaces have developed.

Puffiness

In extreme cases a "puffy" tomato resembles a bell pepper, having normal outer walls and a hollow inside. In most cases only one of the seed cavities will be empty.

Puffiness appears most frequently on early harvests. It is caused by any condition which interferes with normal pollination—temperatures above 90° F. or below 58° F., low light, excessive nitrogen, or heavy rainfall, particularly when accompanied by low temperatures. It can also be caused by chemical sprays intended to increase early yields.

There are no reliable controls.

Sunscald

The USDA has this to say:

"Sunscald is most common on immature, green fruit. At first a yellow or white patch appears on the side of the fruit toward the sun. This spot may merely remain yellow as the fruit ripens, but frequently the tissues are more severely damaged and a blister-like area develops. Later this shrinks and forms a large, flattened, grayish-white spot with a dry paperlike surface."

Varieties differ in resistance to sunscald. Those with poor foliage cover, or varieties susceptible to sunscald damage, can be grown in a cage rather than allowed to sprawl and expose their fruits to the sun.

Insect troubles

The actual damage done by insects to tomato plants and fruits is minimal compared to the damage they do by

Some physiological disorders

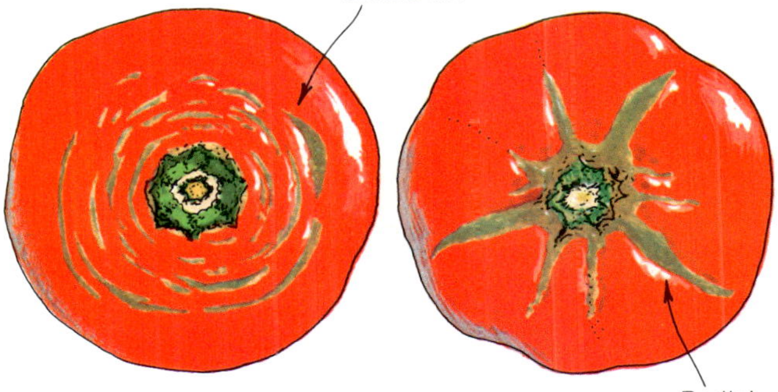

GROWTH CRACKS

CATFACE

SUNSCALD

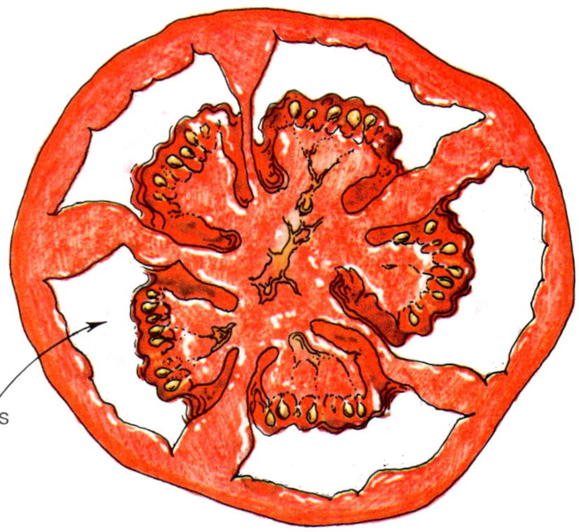

PUFFINESS

BLOSSCM-END ROT

ROOT-KNOT NEMATODES

spreading diseases from plant to plant. It is for this reason that control of insects is important. Also recommended is the removal of weeds from areas where tomatoes are grown; they are breeding places and hiding places for many insects.

Whiteflies

No pest can be as frustrating as the whitefly. Seldom will there be only one or two whiteflies; disturb a leaf on which they are feeding, and they fly from the leaf in a cloud of small white wings.

One reason they are difficult to control is that all forms of the whitefly, from egg, to larvae, to adult are present on the leaf surface.

This makes timing and frequency of insecticide applications the most important part of controlling whiteflies. USDA research has found that an application of Malathion applied four times at ten-day intervals reduced whitefly numbers 26.5% while the same material applied eight times at 5-day intervals caused a 99.7% reduction. Make sure that the product you use has tomatoes on the label. Make sure plants receive full coverage when you spray, and read all product label instructions carefully.

Insects and diseases

Pest or disease	Symptoms	Control
Aphids (plant lice) Attack plants from the seedling stage, through the growing season.	Suck sap; weaken the plant. May deform and stunt the infected terminals and fruit.	Spray with Malathion or Diazinon. Follow instructions on label.
1) **Blister beetles** May attack foliage during midsummer. 2) **Flea beetles** Attack the foliage of young plants until first fruit set. 3) **Colorado potato beetle** General feeder throughout the growing season.	1) Rarely destroy significant amount of foliage. Severe irritation or blistering may occur if beetles are crushed and come into contact with skin. 2) Leaves look as if they had been shot through with tiny holes. 3) Humpbacked larvae hatch; chew on leaves.	1) When harvesting fruit, be careful how you remove Blister beetles—don't crush them. Spray or dust with Sevin. 2) Spray or dust with Sevin. Repeat as directed. 3) Spray or dust with Sevin when damage is first noticed.
Cutworms Most destructive to early season plantings.	Hide in soil during the day, feed at night. Cut off young transplants near the soil surface.	Some gardeners clean them out of the soil before planting with a soil application of Diazinon dust. Other gardeners guard the plant base from cutworms with a waxpaper collar 3 inches high (2 inches above ground and 1 inch beneath the surface).
Hornworms Generally seen on mature foliage.	Eat leaves and fruit, leaving only midribs.	Handpicking is successful for a few plants, or control with Sevin or Diazinon.
Leafhoppers Damage throughout the season.	Leafhoppers suck the sap from the leaves, causing curling and brown margins on the leaves.	Spray or dust with Sevin. Follow label instructions.
Leafminers Seldom a serious problem. Can occur at any time.	Larvae make long, slender, winding white tunnels in the leaves which may result in leaf loss. Tunnels provide entrance sites for diseases.	Spray with Diazinon. Apply as directed.
Tomato fruitworm A pest the whole season long.	Eats into the fruits.	Control difficult after worms are ½-inch long. Control worms when they are small with Sevin.
Early blight Cause: *Alternaria solani,* a fungus. First noticed on the leaves and eventually on the fruit as it matures.	Small, irregular, brown dead spots appear on lower, older leaves. Spots enlarge to ½-inch in diameter in "bullseye" patterns. Leaf tissue around brown spots yellows. With severe spotting, entire leaf will turn yellow. Greatest injury occurs as fruit begins to mature. Older fruits may show dark, leathery, sunken spots, which may be large, and in "bullseye" markings. Dark dry decay may extend into fruit.	Spray or dust plants every 7 to 10 days with Maneb, Zineb or Captan. Repeat after rains to keep plants protected.
Late blight Cause: *Phytophthora infestans,* a fungus. Can be serious during long periods of muggy weather with cool nights and moderately warm days. Hot dry weather checks the advance of the fungus.	Greasy, black areas on leaf margins, spreading over the whole leaf. Fine gray mold on underside of leaf during wet periods. Green and ripe fruits get corky brown on surfaces, develop an "orange peel' texture.	Spray or dust plants every 7 to 10 days with Maneb, Zineb or Captan. Repeat after rains to keep plants protected.
Gray leaf spot Cause: *Stemphylium solani,* a fungus. Most common in warm, moist weather.	Brown spots in circular shapes with gray centers on leaves, later may crack and tear. Stems sometimes spot. Infection begins on older leaves. Leaves yellow, wither and drop. Almost complete defoliation may occur.	Plant disease-resistant varieties. Remove perennial weeds which may harbor disease and insects which spread disease. Follow label directions for fungicide applications.
Tobacco mosaic virus Spreads easily by handling of diseased plants and then healthy plants. Also spread by insects.	Foliage is dwarfed, and mottled in yellowish and dark green spots, and streaks, and develops rough texture. Mottled areas may brown and die. Number of blossoms may be normal, but few fruit are borne. Fruit may be streaked and deformed.	Before handling plants wash hands. Avoid touching plants or flowers after handling tobacco. Destroy infected plants. Use insecticide early and regularly as directed to prevent virus spreading by insects. Remove perennial weeds from the area.
Bacterial spot Cause: *Xanthomonas vesicatoria,* a bacteria.	Small, water-soaked spots enclosing black pustules appear on fruits. Spots sink; scabs form. Infected flower stems cause blossom drop. Seedling stems may be spotted. On mature plants the older leaves are mainly affected. Leaves show irregular dark, greasy spots. Spot centers dry out and tear. Defoliation may be severe.	Use commercial disease-free seed.

Some tomato diseases

. . . and some insects

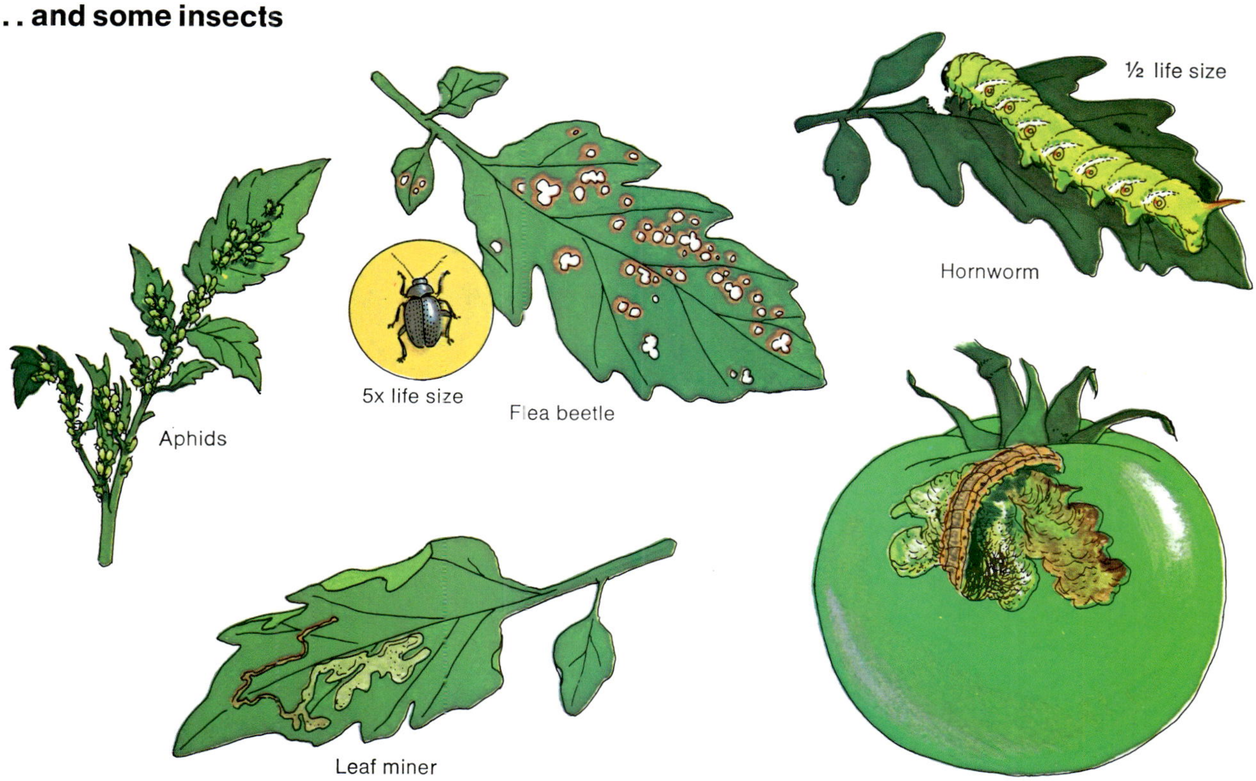

Let's set the record straight

As this is being written there is much talk and a very thorough investigation going on about the safety factor in canning tomatoes.

The U.S.D.A., with well-documented facts, is attempting to nail down the false rumors and clear up old myths.

The rumors have swirled about the acidity of tomatoes and the dangers in canning them. You may have heard "You can be poisoned by home-canned tomatoes." "Many of the newly introduced varieties are too low in acidity to can in the usual way." "All California varieties are lower in acidity than Eastern varieties." "The small tomatoes should not be used in canning."

G. M. Sapers of the United States Department of Agriculture at the Eastern Regional Research Center in Philadelphia, Pennsylvania put it this way:

"Considerable concern has been expressed by the public over the risk of botulism from home-canned tomatoes. This has been engendered in part by recent magazine articles and news items, some of which unfortunately contain inaccurate and misleading information. It has been stated, for example, that 'new strains of tomatoes,' including 'pale yellow-orange tomatoes as well as small cherry or patio ones,' contain insufficient acid to prevent the growth of *Clostridium botulinum.* Such statements may also imply that the consumer of home-canned tomatoes faces a significant danger.

"The seriousness of botulism should never be understated. However, we believe that the risk to home canners of tomatoes is very small. Data compiled by the Center for Disease Control, U.S. Department of Health, Education and Welfare, show only a handful of cases of botulism due to home-canned tomatoes in recent years, none fatal. Furthermore, these documented cases were not associated with low-acid tomatoes."

The U.S.D.A. has collected data on 356 tomato varieties from reports from every region in the U.S.A. These reports were in addition to their continuing study of varieties at Doylestown, PA, and Beltsville, MD. It is from this data that we quote the following pH readings.

Myth

For more years than we can measure, it has been assumed by catalog writers and garden writers that the mild-tasting orange tomato varieties were low acid (high pH).

The yellow and orange varieties have been described in seed catalogs as: "low in acid," "low acid content," "lower in acid."

Fact

The varieties are *not* low in acid. These varieties are normal or high in acidity.

Jubilee	**4.22**	**Yellow Pear**	**4.40**
Sunray	**4.21**	**Yellow Plum**	**4.31**

You can't always taste acidity. Note on the chart at right that honey is high in acidity.

Myth

Our favorite catalog says this about the pink Oxheart: "Firm red flesh is non-acid, instead of being flat tasting it has a pleasant flavor." From down Arkansas way we hear praise of the mild pink nonacid 'Traveler.'

Fact

Pink skin is not a sign of low acidity. Many of the varieties are mild in flavor, but they equal the red ones in acidity.

Oxheart	**4.3**	**Pink Ponderosa**	**4.37**
Traveler	**4.3**		

Myth

White varieties are described in the catalogs as: "creamy-white skin and flesh, fruit free of the usual acids," "extremely mild, and can be eaten by people who avoid tomatoes on account of the acidity," "contains less acid than any other variety."

Fact

These natural assumptions have been made because of the color of the skin and the mild, sweet flavor of the fruit. Actually, the white tomatoes are just as acid as the red tomatoes. Here are the pH ratings on the white varieties:

White Queen	**4.21**	**Snowball**	**4.16**
White Beauty	**4.3**		

Rumor

The small tomatoes and the 'Patio' variety are less acid than the large tomatoes.

Fact

The small tomatoes and the 'Patio' variety are in the same pH range as the large-fruited varieties. Witness:

Basket Pak	**4.3**	**Red Cherry**	**4.4**
Patio Hybrid	**4.3**	**Small Fry**	**4.15**
Presto	**4.1**	**Stakeless**	**4.35**

Rumor

Tomatoes in California have a much higher pH rating than tomatoes grown in the Eastern states. Some of the California processing varieties are higher in pH, however, all available varieties are in the normal pH range.

Fact

There are differences between states, but the variation follows no regional pattern.

Iowa	**4.57**	**Florida**	**4.38**
New York	**4.44**	**Mississippi**	**4.39**
California	**4.43**	**Washington**	**4.38**
Louisiana	**4.42**	**Texas**	**4.38**
Oregon	**4.42**	**Michigan**	**4.38**
Indiana	**4.35**	**Pennsylvania**	**4.35**

"There is probably as much variation in the pH and acidity of tomatoes due to climate, soil, cultural practices and ripeness as to variety differences."—U.S.D.A.

Rumor

New varieties are higher in pH than old varieties.

Fact

New varieties may be lower or higher in pH than old varieties.

Old		New	
Valiant	**4.37**	**Fantastic**	**4.26**
Oxheart	**4.30**	**Early Girl**	**4.14**
Earliana	**4.37**	**Spring Set**	**4.40**
Rutgers	**4.29**	**Spring Giant**	**4.18**

Tomato pH—Past and Present

Year of Introduction	Number of Varieties or Breeding Lines	Mean pH
Before 1950	49	4.29
1950-1959	26	4.34
1960-1969	73	4.35
1970-1976	96	4.34

The U.S.D.A. is continuing to study the safety factors related to individual varieties, as well as the effect of various methods of acidulation. The U.S.D.A. concludes that:

"The home canner should not be overly concerned about hazards associated with the selection of specific tomato varieties for home canning. It is far more important for the home canner to select tomatoes which are not over-ripe, to follow the recommendations of reliable canning guides explicitly, and to destroy (without tasting) any home-canned product which appears abnormal in any way. We especially recommend that the home canner obtain Home and Garden Bulletin No. 8, "Home Canning of Fruits and Vegetables," which is available for 45¢ from the Superintendent of Documents, U.S. Government Printing Office, Washington, DC 20402."

(See page 95 for present recommendations for the amount of citric acid or lemon juice needed to reduce pH levels in tomatoes.)

What is pH

The following definition of pH is from Webster's New Collegiate Dictionary:

"pH: The negative logarithm of the effective hydrogen-ion concentration of hydrogen-ion activity in gram equivalents per liter used in expressing both acidity and alkalinity on a scale whose values run from 0 to 14 with 7 representing neutrality, numbers less than 7 increasing acidity, and numbers greater than 7 increasing alkalinity; also, the condition represented by such a number."

Note carefully that the definition says, ". . . numbers less than 7 increasing acidity, and numbers greater than 7 increasing alkalinity . . ."

In other words, when you say it has a high acidity, it has a low pH, and when it has a high alkalinity, it has a high pH.

There are 2 measurements of acidity used in tomatoes: 'titratable acidity,' and 'pH.'

Here is the chart of pH values for various foods:

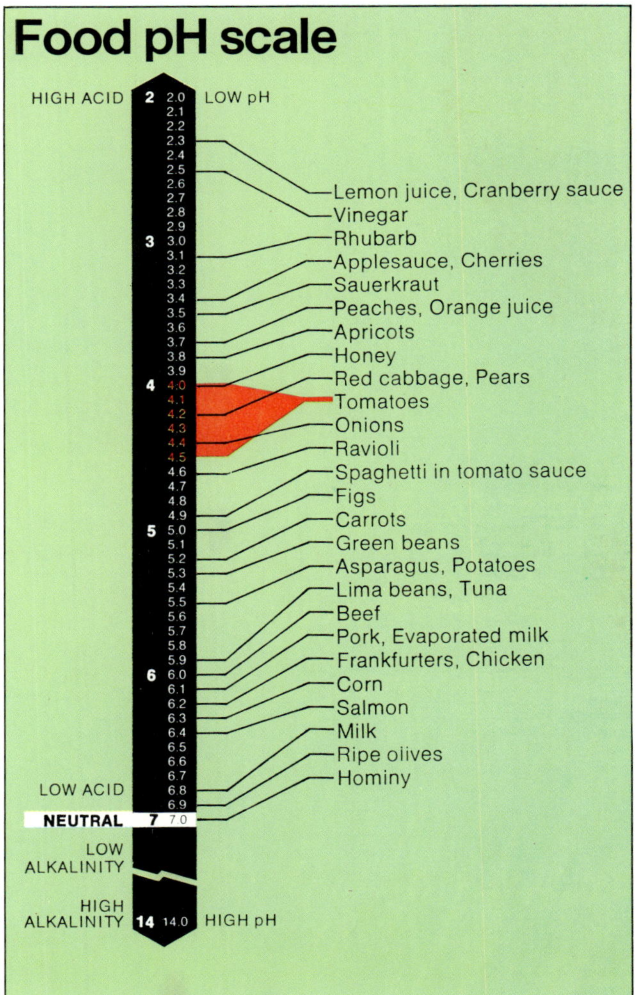

All in the family

A collection of common and not so common garden plants, closely related to the tomato which may find their way into your kitchen including: several potato varieties, some sweet and hot peppers, the eggplant, and a couple of surprises.

The tomato is a member of the Nightshade family, a surprisingly varied group that includes several notoriously poisonous plants, plus such well-known flowers as the petunia, nicotiana, browallia, nierembergia, schizanthus, and other color producers. It also embraces several popular vegetables that have earned a place in the modern garden.

Peppers

The pepper is as temperamental as the tomato. It wants plenty of warm weather but in extremely high summer heat will not set fruit. Blossom drop can be expected when day temperatures reach 90°. As with tomatoes, *night* temperature is also important. Fruit setting is poor when night temperatures are below 55° or above 75°. Plants should not be set out until weather is consistently warm. Early planting followed by cool weather stunts growth, from which the plant never fully recovers.

Another complaint heard about both tomatoes and peppers, is "all vine and no fruit." Too much fertilizer in the soil when the plants are set out is often the cause. Many gardeners wait until fruit has set before applying fertilizer.

The decorative peppers

Many of the peppers lead a double life—as glamorous vegetables, and as ornamentals in the flower border or as container plants (in tubs, boxes, or large pots).

◁

These colorful peppers, dried and fresh, hot and sweet, are as varied to the eye as they are to the taste. The pepper and tomato are closely related in structure, origin and often their use together in the kitchen.

Their beauty in flower, foliage, and fruit earns them a place in the garden or on the patio, even if a pepper is never eaten.

There are many types and numerous varieties of peppers. The sweet-fleshed, big bells—ideally suited for stuffing—are represented by 'Bell Boy,' 'California Wonder,' 'Yolo Wonder,' and 'Florida Resistant.' All of these are generally harvested in their crisp, green stage, but all turn red at maturity. Bell peppers that remain yellow when fully ripe are 'Golden Calwonder' and 'Golden Bell.'

All sweet peppers are not bell peppers. For example, the variety 'Sweet Banana,' with its long, yellow, slender fruit is great for salads or frying. 'Italian Sweet' is an early green pepper.

The heart-shaped pimiento is highly ornamental in a container, and is usually allowed to ripen into a deep red.

To those who know their peppers, the garden should give space to the moderately hot 'Anaheim Chili.' When boiled, skinned, and stuffed with meat or cheese, you have *Chili rellenos,* a tasty Mexican dish.

In our plantings, some of the most attractive plants grown in containers have been hot peppers. 'Hungarian Wax,' an early hot pickling variety about 5 inches long, slender, with a blunt end. Plant grows 12 to 15 inches high and is very prolific under favorable conditions. Fruit is a uniform clear yellow, maturing to bright red. 'Red Cherry' produces small, round fruits held upright on strong, 20-inch-tall plants. 'Long Thin Cayenne,' its name describing the shape of the fruit, but giving no indication of its flavor: very hot. 'Serrano Chili,' small, club-shaped fruits, extremely hot.

Eggplant

Don't try to rush the season with eggplant. Set out plants only after the weather has warmed up—when daytime temperatures are in the 70° range. The plant is more susceptible to cold damage than the tomato. You'll get a higher quality fruit when development is rapid and uninterrupted.

Try the new Oriental varieties. They're smaller than the eggplants you see in the supermarket, but require less total heat for fruit production.

Eggplant makes an attractive container plant, and does very well in a 5-gallon can or box. Place the container in one of the warmer spots in your garden for the best results.

Pick when immature

Eggplant should be used just after harvest as a *fresh* vegetable, not held in storage. For best eating quality, harvest while it is immature and the flesh is firm. In mature fruits the seeds become brown, the seed coats harden, and the flesh softens.

Since you can't see inside the fruit to check the stage of development, it's difficult to tell when they're just right. Harvest the fruits regularly as they reach a usable size.

Harvest oval varieties when 4 to 6 inches long and 3 to 4 inches wide; slim types, 5 to 6 inches long and 2 to 3 inches in diameter. When all fruits are harvested at the immature stage, plants usually set more fruits than when some are allowed to go to full maturity.

Ten days after harvest is as long as you want to hold fruit, before they lose moisture, shrivel, and get flabby. Keep them in a plastic bag to retard moisture loss. Eggplant is subject to chilling injury at temperatures lower than 50°. Store in a cool place rather than the refrigerator.

Husk tomato

The husk tomato is listed in seed catalogs as 'Ground Cherry' and 'Poha Berry.' The ornamental Chinese lantern plant is a close relative. The plant grows from 1½ to 2½ feet tall with oval or heart-shaped fuzzy leaves growing in pairs on the stem. The fruits, about the same size as the cherry tomato, are produced inside a paperlike husk. When ripe, the husks turn brown and the fruit drops from the plant. If left in the husks, they will keep for several weeks.

The husk tomato has a pleasing and distinctive flavor and is used raw in cocktails or for a dessert. It is frequently used in pies, or as a cooked sauce on cakes and puddings, but the favorite way of using pohas is in jam.

Tomatillo

The tomatillo is an annual, growing 3 to 4 feet high. Leaves are long and oval shaped, deeply notched. Fruit is smooth and sticky, either green or purplish in color, and entirely enclosed in a thin husk. Fruit is 1 to 2 inches in diameter, growing to about the size and shape of a walnut.

A close relative of the husk tomato or ground cherry. The tomatillo has a flavor that is tart, rather than sweet like the husk tomato, tasting something like green apples.

It can be used in many ways in Mexican cooking. Raw or cooked, it gives sauces a rich, distinctive flavor. It is used fresh in salads, in tacos, and sandwiches.

Grow tomatillo the same way as tomatoes. Seed sown in peat pots will germinate in about 5 days and be ready to transplant in 2 to 3 weeks.

Tomatillos are harvested according to use. For highest quality fruit, harvest fresh when the husks change color from green to tan. For cooking, harvest at any time. Left on the vine, they become yellow and mild.

Tomatillos can be stored for months. Some gardeners store the fruit on the vine; some spread out the picked fruit, still in their husks, in a cool place with good air circulation.

Irish potatoes

Measured in pounds or dollar return per square foot of garden space, potatoes earn a place in the smallest garden.

You can start new plants from the potatoes you buy at the market. However, market potatoes might carry plant diseases, and might have been treated to prevent sprouting. In any case, freshly dug potatoes won't sprout until they have had a rest period.

It's best to start with certified seed potatoes, or seed pieces, or "eyes" from garden stores and mail-order seed companies. Good-sized seed pieces increase the chances for a good yield. Cut pieces about 1½ inches square, making sure that there is at least one good eye in each piece.

Planting

The conventional planting method is to set seed pieces, cut side down, 4 inches deep, 12 inches apart in rows 24 to 36 inches apart. The tubers (what you eat) form on many stems rising from the seed piece. The potatoes do not grow in the roots. They form above the seed piece on underground stems. When plants are 5 to 6 inches high, scrape soil from between the rows and hill up the plant (cover stems with soil). Potatoes exposed to light, either in the garden or storage, turn green and become inedible.

Growing

Apply fertilizer in bands at both sides of seed pieces at time of planting. Best method is to make a 3-inch-deep, 6-inch-wide trench. Place seed pieces in a row in the center and work in fertilizer 1 to 2 inches deep at the edges of the trench.

A steady supply of moisture is necessary. If soil dries out after tubers begin to form, a second growth starts when soil becomes moist again. The result is knobby potatoes or multiples.

Alternate wet and dry conditions will also cause "hollow heart" (cavities near the center of the tuber).

Cool nights are also needed for good tuber formation. The complaint of "all tops and few potatoes" comes from gardeners who ignore the best planting dates in warm-summer areas.

Picking rather than digging

An old-time method of growing potatoes that you pick rather than dig is this. Set seeds in a wide trench 3 inches deep and as stems grow, build up a covering of straw or pine needles or any material that will protect them from the sun. Cover the material with ½ inch of soil if wind is a problem. The potatoes will form almost at ground level and can be picked up by pulling back the straw. Early potatoes can be picked when tops begin to flower. They will reach full size when the tops die down.

Early-to-medium varieties (90 to 110 days)

'Cobbler.'
White. Nearly round with pronounced eyes. Fairly mealy. Keeps well.

'Superior.'
White, oval, medium size. Slightly russeted. Used for baking, boiling. Some scab resistance.

'La Soda.'
Red. Round. Boiling and potato salad. Does not store well. Good yields.

Late varieties (110 to 140 days)

'Katahdin.'
White. Round, smooth-skinned, with shallow eyes. Baking, boiling. Resistance to scab, drought, verticillium.

'Kennebec.'
White. Block-shaped. Good for frying and hash browns. Stores moderately well. Resistance to late blight, some scab.

'Red Pontiac.'
Tends to become over-sized with abundant rainfall. Fair table quality. Stores well.

'Sebago.'
White. Oval to oblong, shallow eyes. Fair resistance to scab and late blight, resistance to Yellow Dwarf disease.

Tire-raised bed

To concentrate the growing of potatoes, gardeners have made use of old tires to hold the soil and frame the potatoes. Potato seed pieces are planted just below ground level, and as the vines grow up they are mulched with a lightweight soil mix. On the theory that potatoes will form along the above-ground stems, additional tires are added as the plants grow. From the reports we have received, the optimum production is about 10 pounds per tire. We have fallen short of this yield in our trials.

Tree tomatoes

Two types of tree tomatoes are available to the home gardener. The domestic tree tomato, listed in seed company catalogs as 'Climbing Tomato,' 'Trip-L-Crop,' as well as 'Tree Tomato,' grows as a vigorous, indeterminate vine, capable of spreading to 18 feet. It produces large (a pound or more) smooth-skinned fruit. It is heavy-stemmed with potato-type leaves.

The subtropical tree tomato *(Cyphomandra betacea)*, commonly called 'Tamarillo,' has been enthusiastically promoted as an outdoor-indoor tree producing hundreds of tomatoes over a 5-month period. In all but a subtropical climate it should be treated as an indoor plant. We donated the plant we grew to a local kindergarten classroom. It has never borne fruit, (even though it's more than 3 years old). But it is prized for its exotic, elephant-ear-like leaves, and waxy white blossoms. Taken from the classroom in summer, it receives a severe pruning back to a few sticks. When school resumes in fall, the new foliage growth is full and fresh.

The fruit from commercial orchards in New Zealand is now being imported to California. The fruit is about the size of a large AA egg, tough-skinned, and quite tart.

For detailed information on how to use the fruit, send a self-addressed, stamped envelope to Frieda's Finest, Produce Specialties Inc.,
732 Market Court, Los Angeles,
CA 90021.

Seed sources

Husk tomato (4) (9) (12) (16) (20).

Tree tomato—domestic (4) (15) (16) (20) (21) (22) (31).

Tree tomato—imported (31).

Potato—check your garden store for local supply of certified seed potatoes. Availability: (9) (12) (16) (20) (50).

Eggplant—many varieties widely distributed. For Oriental varieties check (44) (45) (48) (*56) (*57).

Peppers—widely available. For unusual and hot pepper varieties, tomatillo, etc. check (52).

Potatoes produce heavy crops in old tire "raised beds."

Tomatillo, in and out of its husk, revealing its solid flesh.

Branch of the husk tomato shows its resemblance to the Chinese lantern plant.

From vine to kitchen

To the tomato "farmer" there's no such thing as too many tomatoes—especially when there's a good cook in the kitchen. The good farmer plans a succession of harvests with the kitchen in mind: from the first eagerly awaited early varieties, to the mid-season canners, to the last green tomatoes when frost puts an end to the growing season.

For some gardeners the pleasure of growing the perfect tomato reaches its peak with a midsummer trip to the garden, salt shaker in hand, and the picking of a perfect, vine-ripe tomato. This juicy, sun-warmed bit of perfection marks the ultimate achievement for these tomato lovers.

For other gardeners, the pleasure of growing the perfect tomato is only a beginning. They know that many delicious endings are in store for that tomato, once a good cook takes over.

During the course of our tomato trials, we were fortunate to have the talents of a chef with an inherited love for tomatoes. With her, we followed our tomatoes from the garden, through the kitchen and on to the table.

Here we surrender this book to our good friend and chef, Annette Fabri. "My first association with tomatoes was at a tender age. I remember, as a child, the jar of tomato sauce in grandmother's icebox. I remember it well, because the sauce was used so often. Hers was the theory that almost any dish could be improved with a little tomato flavor or color. Her belief was passed on to my mother, and it is mine today.

"Although I have favorite herbs and spices that I use for seasoning, I am never without a jar of tomato sauce. When tomatoes are in season, I prepare a fresh tomato sauce, lightly seasoned with onion, bell pepper, and herbs. This will keep for about a week in my refrigerator. During the winter I substitute pints of home-canned tomato sauce for the fresh.

"I am fortunate in having a husband and son who share my fondness for tomatoes. We use them in everything from scrambled eggs to basting sauces for broiled and roasted meats —to biscuit and bread dough.

"The versatility of the tomato has never ceased to amaze me. I have discovered that it is possible to take the tomato, in one form or another, through the entire sequence of a meal —from appetizer to dessert.

"Among the recipes that follow are some oldtime favorites, as well as some new recipes which use the tomato in less than traditional ways; from the first green tomatoes to the last of the abundant harvest."

The tomato—the star performer in the garden and in the kitchen, throughout the harvest seasons.

Sauces and condiments

Many recipes in the following pages call for prepared products such as tomato sauce, stewed tomatoes, tomato juice, catsup, and tomato paste. These products are, of course, available at grocery stores and supermarkets, but by preparing them in my kitchen, I have the freedom to add my own touches. The following are some of my basic recipes for these products.

Tomato Sauce

This is the sauce that I always have on hand in a jar in my refrigerator because it has a multitude of uses: on steaks, chops, and chicken before broiling; on leg of lamb, beef or pork roasts while roasting, to give a rich brown color. I also pour it over stuffed bell peppers, meat loaf, and baked fish. Combine it with scrambled eggs, fresh or frozen vegetables, baked rice, and include it in an endless number of casseroles. I use it when making biscuit and bread dough in place of some of the liquid normally called for, and as a base for Creole dishes, spaghetti sauce, barbecue sauces, soups, and salad dressings.

1 onion, chopped
2 tablespoons olive or salad oil
1 small clove garlic, crushed
2 tablespoons chopped parsley
¼ cup chopped green bell pepper
3 cups peeled and chopped fresh tomatoes (or 3 cups canned, whole tomatoes)
1 teaspoon salt
½ teaspoon pepper
½ teaspoon dried oregano
1 bay leaf

1. Cook onion in oil until lightly browned. Add garlic, parsley, and bell pepper and stir until limp.

2. Add tomatoes, salt, pepper, oregano, and bay leaf. Simmer, uncovered, until thickened, about 1 hour.

3. Remove bay leaf and store sauce in covered container in refrigerator. Makes 2 cups.

Stewed Tomatoes

6 medium-size ripe tomatoes
½ cup minced onion
¼ cup chopped green pepper
¼ cup chopped celery
2 tablespoons salt
½ teaspoon pepper
1 teaspoon white sugar
1 tablespoon chopped parsley
1 tablespoon butter
Buttered bread crumbs (optional)

1. Peel, core, and quarter tomatoes. Place in a saucepan together with onion, green pepper, and celery. Sprinkle with salt, pepper, sugar, and parsley.

2. Bring to a boil, cover, and simmer 10 minutes.

To peel a tomato: 1. Cover with boiling water for 10 seconds.

2. Immerse in ice water until cooled.

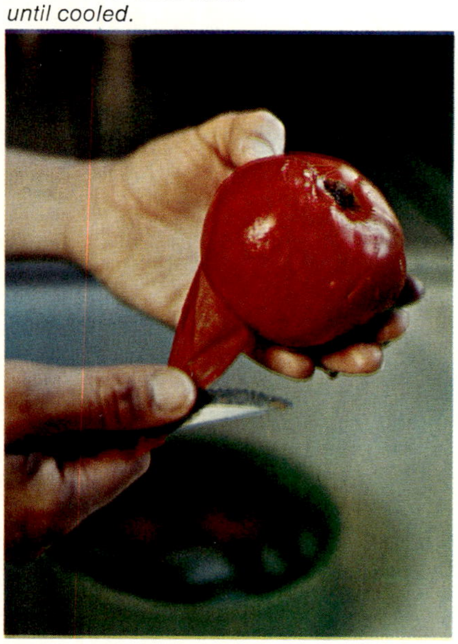

3. Remove peel with a serrated knife.

3. For serving as an individual dish, dot with butter and sprinkle with buttered crumbs. Makes 4 servings. Or, use in combination with other ingredients in casseroles and stews.

Tomato Paste

The best tomato paste is made with the Italian, plum-type tomatoes. It has many uses and a small amount will go a long way. I use it to add color to otherwise pallid dishes, and for thickening tomato sauces.

1. Peel and chop 4 pounds of tomatoes. Measure and add 1 teaspoon salt for every pint of chopped tomatoes.

2. Place in a large kettle and simmer over low heat for about 1 hour. Stir often to keep from sticking.

3. Remove from heat and put through a food mill or fine sieve.

4. Return to kettle and continue to cook very slowly, until paste holds its shape on a spoon; takes about 2 hours. Stir occasionally to prevent sticking.

5. Pour into sterilized ½-pint jars, filling to within ½ inch from top. Seal and process in boiling water bath for 30 minutes. (See pages 90-92.)

Cocktail Sauce

An excellent sauce for dunking small sausages or other hors d'oeuvres, or for garnishing shellfish.

¾ cup chili sauce or catsup
3 tablespoons lemon juice
3 tablespoons prepared horseradish
2 teaspoons Worcestershire sauce
1 teaspoon grated onion
Dash Tabasco sauce
Salt and pepper

1. Combine all ingredients. Add salt and pepper to taste. Chill.

2. Serve with seafood cocktails or as a dunking sauce. Makes 1½ cups.

Tomato Catsup

There are many recipes for tomato catsup. The difference in most lies in the amount of spices used. The following is a typical recipe; the spices may be adjusted to suit one's own taste.

6 quarts (9 pounds) fresh, ripe tomatoes
3 medium onions, peeled and sliced
2 large, sweet red peppers, seeded and sliced
1 stalk celery, strings removed
2 cloves garlic, crushed
1 cup cider vinegar
1 teaspoon dry mustard
1 cup sugar
1 teaspoon paprika
1 tablespoon salt
1 teaspoon whole peppercorns
1 teaspoon whole cloves
2 teaspoons whole allspice
1 stick (2½ inches) cinnamon, broken

1. Remove stem and core from tomatoes; quarter. Place in a large kettle, together with onions, celery, and red peppers. Cook, uncovered, until soft, about 30 minutes.

2. Put through a food mill. Return pulp to kettle and add garlic, vinegar, mustard, sugar, paprika, and salt.

3. Tie whole spices in a cheesecloth bag and add to kettle.

4. Bring to a boil, lower heat, and simmer, uncovered, one hour or more, or until mixture is quite thick.

5. Remove spice bag. Pour into sterilized jars. Seal. Process in boiling water bath 5 minutes. (See pages 90-92.) Makes about 2 pints.

Chili Sauce

8 pounds ripe, juicy tomatoes
3 sweet red peppers
3 green bell peppers
1 stalk celery
2 cups chopped onion
2 cloves garlic, crushed
1 teaspoon each: whole allspice, mustard seed, whole cloves
1 cup light brown sugar
2 tablespoons salt
1 teaspoon each: fresh ground black pepper, dry mustard
2 dried, hot red peppers, crushed
1½ cups cider vinegar

1. Scald, peel, core, and chop the tomatoes. Chop peppers and celery. Add, with the onion and garlic, to the tomatoes. Bring to a boil and simmer for 45 minutes.

2. Tie whole spices in a cheesecloth bag and add to tomato mixture. Add sugar, salt, black pepper, mustard, and hot peppers. Boil, uncovered, until thick.

3. Add vinegar and boil the sauce until it has thickened to the correct consistency. Discard spice bag.

4. Pour into sterilized jars. Seal. Process in boiling water bath for 5 minutes. (See pages 90-92.) Makes 8 pints.

All-Purpose Barbecue Sauce

This is an excellent barbecue sauce to keep on hand. It makes a large amount, but keeps for months in the refrigerator. We use it for all meats; it is especially good on poultry.

½ cup chopped onion
¼ cup olive oil
2 cups sauterne wine
1 cup soy sauce
2 cups catsup
½ cup prepared mustard
¼ cup sugar
Salt and pepper to taste
¼ cup chopped parsley

1. Cook onion in olive oil for 5 minutes.

2. Add wine and simmer until reduced by half.

3. Add remaining ingredients, except parsley, and simmer for 20 minutes. Do not boil at any time.

4. Remove from heat and add parsley. Adjust seasoning to taste.

5. Pour into a jar and keep refrigerated until needed. Makes 1 quart.

6. To use, brush on meat about 15 minutes before barbecuing. Use as a basting sauce while meat is cooking.

Fresh Tomato Barbecue Sauce

This is a basic barbecue sauce that can be used on all meats. Variations are possible depending on taste preferences.

½ cup chopped onion
1 clove garlic, crushed
½ cup salad oil
2 cups peeled, seeded, and chopped tomatoes
2 tablespoons lemon juice
⅛ teaspoon Tabasco sauce
1 teaspoon dry mustard
2 tablespoons brown sugar
1 teaspoon dried oregano
½ cup dry red wine
Salt and pepper to taste

1. Sauté onion and garlic in oil. Add tomatoes and simmer until soft.

2. Add remaining ingredients, except wine, and simmer gently for 30 minutes, or until thickened.

3. Add the wine and heat thoroughly. Season with salt and pepper.

4. Cool slightly. Pour into blender and whirl until smooth. Store in refrigerator in covered container. Makes 3 cups.

Variations:

✓Substitute dried thyme or basil for the oregano.

✓Stir in a bit of chili sauce just before sauce is ready to serve.

✓Substitute an equal amount of white sugar for brown.

✓Sauté half a green pepper, chopped, with the onion and garlic.

Tomato Juice

Select tomatoes that are ripe and juicy. Wash; remove stems and cores. Cut into quarters. Simmer in a large kettle until soft, stirring often. Put through a food mill or fine sieve to remove peel and seeds. Season to taste.

Refrigerate small amounts in covered container for up to one week.

Larger amounts should be processed for canning, as follows: add one teaspoon salt to each quart of juice. Reheat to boiling and pour immediately into sterilized jars, filling to within ½ inch of top. Seal, screw bands firmly tight. Process in boiling water bath for 10 minutes. (See pages 90-92.)

Tomato-Vegetable Cocktail

2 cups tomato juice, canned or fresh
1 carrot, cut up
¼ small onion
Few celery leaves
Few parsley sprigs
1 cup crushed ice
Dash Worcestershire sauce
¼ teaspoon bottled horseradish

1. Combine all ingredients in a blender.

2. Cover and whirl until blended and smooth.

3. Serve immediately. Serves 4.

Tomato Juice Cocktail

3 cups tomato juice, fresh or canned
⅛ teaspoon pepper
¾ teaspoon salt
¼ teaspoon sugar
1 tablespoon lemon juice
4 drops Tabasco sauce

1. Combine all ingredients in a large pitcher. Chill thoroughly.

2. Stir well and serve in chilled glasses—with lemon wedges, if desired. Serves 6.

Spicy Tomato Juice Cocktail

1 cucumber, peeled and grated
6 cups tomato juice ,fresh or canned
3 green onions, finely chopped
3 tablespoons lemon juice
Dash Tabasco sauce
1 tablespoon Worcestershire sauce
1 tablespoon bottled horseradish
Salt and pepper to taste

1. Combine all ingredients in a large pitcher and chill at least 2 hours.

2. Strain through a coarse sieve.

3. Stir well and pour into chilled glasses. Serves 8.

Tomato Pick-up

We encountered this fresh, spicy drink in Mexico.

3 cups tomato juice
1 cup orange juice
1 fresh or canned green chili, seeds and vein removed
1 teaspoon sugar
½ teaspoon salt
Juice of 1 small lime
¼ cup chopped white onion

1. Place all ingredients in a blender and whirl until combined.

2. Chill thoroughly.

3. Serve in chilled wine glasses.

A pitcher of chilled tomato juice is a good way to start any day and a sure thirst-quencher on a hot summer day.

Sparkling, shimmering tomato aspic delights both the eye and palate, and adds a colorful touch to any meal.

Soups and salads

A refreshing way to begin a meal is with a bowl of homemade tomato soup or a fresh tomato salad.

Fresh Tomato Aspic

3 tablespoons plain gelatin
½ cup cold water
8 large tomatoes
1 teaspoon sugar
1 bay leaf
1 teaspoon salt
2 small onions, coarsely chopped
1 stalk celery, coarsely chopped
1 teaspoon peppercorns
3 tablespoons lemon juice
⅛ teaspoon Tabasco

1. Sprinkle gelatin in water to soften.

2. Quarter tomatoes, cutting away white core and stem end. You should have about 2 quarts.

3. Place tomatoes in a saucepan and add sugar, bay leaf, salt, onions, celery, and peppercorns. Cook over low heat until tomatoes are soft, about 10 minutes.

4. Strain through a food mill and measure 5 cups of tomato juice. If amount is less than 5 cups, add enough water, canned tomato juice, consommé, or bouillon to make 5 cups.

5. Add softened gelatin to hot tomato juice and stir to dissolve, then add lemon juice, Tabasco, and salt and pepper to taste.

6. Turn into lightly greased 5- to 6-cup mold. Chill until firm.

Variations:

Tomato-Vegetable Aspic

Follow steps 1 through 5 then chill mixture until consistency of unbeaten egg white. Fold in 1 cup shredded cabbage, ½ cup chopped celery, and 1 cup finely chopped green pepper. Turn into 7- to 8-cup mold. Chill until firm.

Other Variations:

Follow directions for Tomato-Vegetable Aspic, but substitute one of the following for the raw vegetables:

✔1½ cups cooked vegetables.

✔1 cup diced cooked chicken and ¼ cup sliced olives.

✔1 cup cooked shrimp.

✔1 cup slivered cooked ham and ¼ cup pickle relish

Fresh Tomato Soup

I am particularly fond of tomato soups, both hot and chilled. We have several favorites and I have always been able to make one to compliment the main course of just about any meal I serve.

I've found that fresh, fully ripe, tomatoes make the best tomato soups. However, when fresh tomatoes are out of season, canned tomatoes, peeled, can be substituted.

½ cup butter
2 tablespoons salad oil
3 pounds ripe tomatoes, peeled, cored, and quartered
2 teaspoons chopped fresh thyme (or ½ teaspoon dried)
2 tablespoons chopped fresh basil (or 1 teaspoon dried)
1 large onion, thinly sliced
3 tablespoons canned tomato paste
¼ cup all-purpose flour
4 cups rich chicken broth
1 cup cream
1 teaspoon sugar
Salt and pepper to taste

1. Heat butter and oil in a heavy skillet. Add the onion and cook until translucent.

2. In a large kettle combine tomatoes, herbs, onion, and tomato paste. Simmer for 10 minutes.

3. Mix the flour with ¾ cup of the chicken broth and stir into the tomato mixture. Add remaining broth and cook 30 minutes, stirring frequently.

4. Pass mixture through a fine sieve or whirl in blender until smooth. Reheat and stir in cream and sugar. Do not boil. Add salt and pepper to taste.

5. Serve in cups or bowls with a pat of butter, if desired. Serves 8.

Cream of Fresh Tomato Soup

2½ cups chopped ripe tomatoes
1 medium onion, sliced
1 bay leaf
1 teaspoon salt
¼ teaspoon pepper
2 tablespoons butter or margarine
2 tablespoons all-purpose flour
2 cups milk
Chopped fresh basil

1. Simmer tomatoes with onion, bay leaf, salt and pepper, for 10 minutes, then strain.

2. Make a white sauce by melting butter and blending with the flour. Gradually add the milk and cook over low heat until slightly thick, stirring constantly.

3. Just before serving, slowly add hot, strained tomatoes to the white sauce, stirring until smooth. Garnish with fresh basil. Serves 4.

Gazpacho

Many versions of this cold Spanish tomato soup recommend that all the ingredients be put through a blender, rather than chopped by hand. I prefer to combine both methods, allowing the flavor of each vegetable to be more distinct and giving the soup a nice texture. A good way to keep this soup icy cold is to freeze cubes of plain tomato juice and add them to the gazpacho just before serving.

1 medium cucumber, peeled, seeded, and chopped
1 large green pepper, seeded and chopped
2 stalks celery, pared to remove any strings, and bias cut
½ cup green onion, chopped
1 garlic clove
2 tablespoons chopped parsley
½ teaspoon dried basil (or 2 tablespoons fresh)
2 cups tomato juice, fresh or canned
4 tablespoons olive oil
3 tablespoons vinegar or lemon juice
½ teaspoon monosodium glutamate
4 large ripe tomatoes, peeled, seeded, and chopped
Dash of Tabasco sauce
Salt and pepper to taste
Condiments: Croutons, minced green onions, chopped bell pepper, chopped hard-boiled egg, sour cream, frozen cubes of tomato juice.

1. Place all ingredients except tomatoes, Tabasco sauce, salt, pepper, and condiments into blender. Whirl only until blended, but not pulverized.

2. Pour into large container and add tomatoes, Tabasco sauce, salt and pepper. Taste and adjust seasoning if necessary.

3. Chill thoroughly.

4. Place condiments in individual bowls, to be passed separately. Serves 6.

Hot Tomato Bouillon

This is a first-course soup that is always welcome on a cold winter day. I like to serve it in large mugs, with a sprinkling of crumbled bacon on top. The horseradish is a nice tangy addition.

1½ cups tomato juice, fresh or canned
1 can condensed beef broth
1 cup water
¼ cup egg pastina (tiny soup macaroni)
½ teaspoon prepared horseradish
2 slices bacon, crisp-cooked, drained, and crumbled

1. In a saucepan, combine tomato juice, beef broth, water, pastina, and horseradish. Bring to a boil and simmer 7 to 10 minutes or until pastina is tender, stirring occasionally.

2. Ladle into large mugs and spoon crumbled bacon atop. Serves 4.

A tureen of chilled Spanish Gazpacho takes center stage in a group of condiments—hard cooked egg, minced green onion, chopped bell pepper, sour cream, croutons, and cubes of frozen tomato juice.

Sherried Tomato Seafood Bisque

I like to use various wines in my cooking and find sherry to be one of the most versatile—it is especially good in this seafood bisque, if added just before serving. I also serve a cruet of sherry along with the soup, should any of my guests wish to add more. This is a hearty soup, which can be made with either crab, shrimp, or lobster. Served with hot garlic-buttered French bread and a tossed green salad, it becomes a meal in itself.

4½ cups peeled tomatoes
2 cups beef broth (or use beef bouillon cubes and water, or canned beef bouillon)
1 cup diced celery
2 small onions, sliced
2 small carrots, sliced
3 tablespoons uncooked rice
2 parsley sprigs
4 whole cloves
6 peppercorns
Small piece of bay leaf
Pinch of ground thyme
2 teaspoons salt
1½ pounds crab meat, shrimp, or lobster, cooked and cut into pieces
2 cups light cream
Sherry to taste
Thin slices of lemon and chopped parsley for garnish
Croutons

1. Put tomatoes, broth, vegetables, rice, and seasonings (tie seasonings in a cheesecloth bag) into a kettle. Bring to a boil, cover, and simmer for 1 hour.

2. Remove from heat and discard cheesecloth bag. Put all ingredients into a blender and whirl until smooth.

3. Just before serving add fish; simmer until heated through.

4. Heat cream slightly and add to tomato mixture. Season to taste.

5. Add a dash of sherry; garnish with lemon slices and chopped parsley; serve at once. Pass croutons and extra sherry. Serves 8.

Creamy Iced Tomato Soup

This is one of the best cold soups I know of, and it's surprisingly easy to prepare. It can be made as much as two days in advance. Stir in sour cream just before serving.

2 large ripe tomatoes, peeled and quartered
1 small clove garlic, crushed
1 teaspoon sugar
1 tablespoon salt
½ teaspoon pepper
2 cups tomato juice, fresh or canned
1 cup dairy sour cream
1 large cucumber

1. Press quartered tomatoes through a sieve into a large bowl.

2. Add garlic, sugar, salt, pepper, and the 2 cups of tomato juice. Cover and refrigerate. (Stop here if you are working ahead.)

3. When ready to serve, taste and add additional seasonings, if necessary. If mixture is too thick, stir in additional tomato juice. Stir in sour cream until blended and smooth.

4. Serve chilled soup in cups or bowls, garnished with cucumber slices. Serves 6.

Tomato-Vegetable Beef Soup

This is a very thick, hearty soup that uses a variety of fresh vegetables. Lima beans, chopped cabbage, diced yellow turnip, cauliflower, or whatever you happen to have on hand can be substituted for the vegetables listed.

1 beef soup bone
1 pound soup beef, cubed
1 tablespoon salt
½ teaspoon peppercorns
1 bay leaf
2 quarts water
1 medium onion, chopped
2 large carrots, diced
3 celery stalks, sliced
1 cup each of cut green beans, corn, green peas, and diced potatoes
2½ cups peeled and quartered tomatoes
1 medium green pepper, diced
Salt and pepper to taste

1. Put soup bone, meat, salt, peppercorns, bay leaf, and 2 quarts of water in a large kettle. Bring to boil; simmer, covered, for 1 hour.

2. Add vegetables; bring to boil and simmer, covered, for 1 hour longer.

3. Remove meat from bone and put meat back in soup.

4. Season with salt and pepper to taste. Makes about 2½ quarts.

Chilled Beef and Tomato Salad

Leftover boiled beef or pot roast are ideal for this chilled salad. Prepare it the day ahead, so all ingredients can marinate thoroughly.

3 medium boiled potatoes, peeled
2 large tomatoes, peeled, seeded, and cut into chunks
3 cups boiled beef, cut into strips
1 tablespoon minced parsley
½ cup Italian dressing
1 teaspoon Worcestershire sauce
2 teaspoons chopped onion
1 teaspoon oregano
Salt and pepper to taste

1. Cut potatoes and tomatoes into medium-size chunks.

2. Mix all ingredients in a large bowl. Season to taste, cover, and chill overnight.

3. Serve on a cold platter, lined with lettuce leaves. Serves 4.

Creamy Iced Tomato Soup (recipe column 1) is kept chilled and ready to serve when immersed in a bowl of ice cubes.

The blender performs an important function in this thick and hearty Sherried Tomato Seafood Bisque.

Fresh tomato slices surround this antipasto tray of olives, sardines, hard-cooked eggs, tuna chunks, anchovy fillets, marinated mushrooms, and artichoke hearts.

Antipasto Salad

Antipasto is a fun salad to make. I start with a border of lettuce, either head lettuce or romaine, and sliced tomatoes. In the center of the platter, I give free rein to my imagination in trying to arrange colorful combinations of food—thinly sliced meats and cheeses for a delightful contrast, salami, hard-cooked eggs, tuna, olives (both green and black), artichoke hearts, anchovy fillets, and even pickled beets and sweet red peppers can be added for color. I like to make up combinations that include favorite delicacies of each of my guests. For a large party I make two or three different antipasto trays. The combinations are endless.

In addition to the above, consider some of the following when preparing an antipasto tray:

✔Marinated mushrooms
✔Pickled oysters
✔Stuffed celery
✔Sardines in olive oil
✔Slices of fresh kohlrabi
✔Small green onions
✔Radish flowerettes
✔Boiled shrimp or crabmeat

Mexican Tomato Salad

This salad can be prepared well ahead of serving time. If fresh coriander is unavailable, the same amount of parsley can be substituted. However, the coriander adds greatly to the distinct South-of-the-Border flavor.

2 pounds medium-size zucchini
4 hard-boiled eggs
4 large, fresh, firm ripe tomatoes, peeled, seeded, and diced
½ cup olive or salad oil
1 teaspoon finely minced green chilies
¼ cup finely minced green onions
1 tablespoon minced coriander
3 tablespoons wine vinegar
Salt and pepper to taste

1. Cut zucchini in half lengthwise, then into ¼-inch pieces. Do not peel. Boil in salted water for 5 minutes. Drain well.

2. Cut eggs into large pieces. Place zucchini, eggs, and tomatoes in salad bowl. Add oil and combine thoroughly.

3. Add chili peppers, green onions, coriander, vinegar, and seasonings. Toss well, cover, and chill till serving time. Serves 6.

Cold Stuffed Tomato Salads

Fresh tomato salads are a real hit for a summer luncheon. I like to cut each tomato in a different way and use a variety of stuffings. I then set the salads out on a buffet table and guests can choose their favorites. If I am serving them as a side dish instead of a main course, I cut the tomatoes in halves or slices, place a ring of bell pepper, ½ inch or more thick on each slice, and fill the ring. The contrast of the filling, with the bright green ring in the middle, and the beautiful red tomato on the bottom is very pretty.

Whether to peel a tomato which is to be stuffed or not, depends a great deal on its appearance. If it is good and ripe, it doesn't need peeling. However, if it appears to be coarse and the color is inconsistent, by all means peel. To prepare tomato cases, first peel the tomatoes, if you choose, then hollow them—a serrated grapefruit spoon greatly speeds the process. Sprinkle lightly with salt, invert and allow to drain for 20 minutes. Chill, then fill with stuffing of your choice. A few suggestions follow:

✔Tiny shrimp which have been marinated in Italian dressing. Drain and sprinkle with parsley.

✔Cottage cheese seasoned with chopped, canned green chilies, pimiento, onion, and a dash of Worcestershire.

✔Guacamole mixed with chunks of fresh avocado.

✔Hard-boiled eggs, mashed and seasoned with mayonnaise and curry powder.

✔Potato or macaroni salad seasoned with onion, green pepper, and hard-boiled eggs.

✔Tuna fish and a garlic-flavored mayonnaise dressing.

✔Chicken salad mixed with crisp celery chunks and a mayonnaise dressing.

✔Cold salmon garnished with cucumber slices and served with lemon wedges.

✔Crab or lobster chunks marinated in a little vinaigrette sauce before filling. Serve additional sauce on the side.

✔Crabmeat and a Louis dressing, served with sliced beets, hard-boiled eggs and asparagus spears.

Cold Stuffed Tomato Salads with a variety of fillings lend themselves to playful slicing.

Cherry tomatoes highlight this Vegetable Rice Ring (recipe below) which is teeming with bits of fresh radish, cucumber, tomato, green pepper, celery, and onion.

Vegetable Rice Ring

This salad is a real time saver for me, even though it looks like it might take a long time to prepare. I cook the rice a day ahead and marinate it overnight in Italian dressing. It's always fun to experiment with different raw vegetables to add to it. Bright little cherry tomatoes in the center of the mold not only look good, but they also add an extra special flavor.

2 cups cooked rice
½ cup Italian salad dressing
½ cup mayonnaise or salad dressing
1 cup sliced radishes
1 medium cucumber, seeded and chopped
2 small tomatoes, peeled, seeded, and chopped
1 medium green pepper, chopped
½ cup chopped celery
¼ cup chopped green onions
Leaf lettuce
1 pint cherry tomatoes

1. Combine cooked rice and Italian salad dressing; cover and chill several hours or overnight.

2. Add mayonnaise to undrained rice mixture and stir until combined.

3. Fold in radishes, cucumber, tomatoes, green pepper, celery, and green onions.

4. Press into 5½-cup ring mold which has been greased lightly with salad oil.

5. Cover; refrigerate at least 2 to 3 hours.

6. To serve, unmold onto lettuce-lined platter and fill center with cherry tomatoes. Makes 8 to 10 servings.

Relishes and side dishes

There are many types of tomato relishes. You can enjoy some fresh, and can others. I like to make fresh relishes during the summer when there are so many fresh vegetables available. They will keep for several days in the refrigerator. Whether fresh or canned, both types are excellent served with hamburgers, hot dogs, and barbecued and broiled meats.

Fresh Tomato Salsa

For a something-different relish or appetizer, serve this fresh tomato salsa and corn chips. The fresh vegetables mixed with chopped green chilies and spices get dinner or cocktails off to an extra spicy start.

4 medium tomatoes, peeled and finely chopped
½ cup finely chopped onion
½ cup finely chopped celery
¼ cup finely chopped green pepper
¼ cup olive oil or salad oil
2 to 3 tablespoons canned green chilies, drained and finely chopped
2 tablespoons red wine vinegar
1 teaspoon mustard seed
1 teaspoon crushed coriander seed
1 teaspoon salt
Dash pepper

1. Combine all ingredients.

2. Cover. Chill several hours or overnight; stir occasionally.

Tomato, Pepper, and Celery Relish

3 medium tomatoes
1 green pepper, minced
½ cup diced celery
1 small onion, minced
1½ teaspoons salt
2 tablespoons each of vinegar and sugar
½ cup cold water
⅛ teaspoon salt

1. Peel and dice tomatoes. Combine with remaining ingredients.

2. Chill for several hours. Drain. Makes 2 cups.

It's relish time when the garden yields a variety of fresh vegetables and tomatoes.

Tomato, Onion, and Cucumber Relish

2 tomatoes, sliced in thin wedges
2 onions, chopped
1 medium cucumber, peeled and sliced
2 tablespoons chopped parsley
½ cup Italian salad dressing
Salt and pepper to taste

1. Place vegetables and parsley in a shallow bowl. Pour dressing over the top and mix lightly.

2. Allow to marinate and chill thoroughly.

3. Drain. Add salt and pepper to taste. Makes 2 cups.

Tomatoes Florentine

This is how I turn spinach into a special event. We like these stuffed tomatoes with just about any kind of roasted meat. If you want to prepare them ahead of time, refrigerate the stuffed tomatoes until ready to cook, and add 10 minutes to the cooking time.

6 medium tomatoes
½ teaspoon salt
12 ounces fresh spinach or 2 packages (10 ounces each) frozen chopped spinach
2 tablespoons butter or margarine
2 tablespoons all-purpose flour
¾ cup milk
2 hard-cooked eggs

1. Cut a slice, ¼ inch thick, off the top of each tomato. Scoop out the insides of tomatoes, leaving a shell about ¼ inch thick. Sprinkle insides with salt and set aside.

2. Cook fresh spinach in small amount of boiling water 5 to 8 minutes. Drain well; chop and return to saucepan. (Or prepare frozen spinach according to package directions; drain well.) Add butter or margarine; stir until melted.

3. Blend together flour and milk; add to spinach. Cook and stir until thickened and bubbly.

4. Sieve one hard-cooked egg yolk and set aside. Chop remaining egg and egg white and add to creamed spinach.

5. Fill tomatoes with spinach mixture. Place in a shallow baking dish. Bake at 375° for 20 minutes.

6. Sprinkle tops of tomatoes with sieved egg yolk just before serving. Serves 6.

Barbecued Tomatoes

Although cherry tomatoes seem like the ideal size for skewering, their tough skin and weak flavor are always a disappointment.

With a sharp knife, remove stems from medium-size, firm, ripe tomatoes. Cut each tomato in half, crosswise. Stab tomatoes with two skewers, brush with oil, and lay skewers over glowing coals. Rotate occasionally, until warmed through. The double-skewer method will keep tomatoes from revolving when turned.

Wine-glazed Tomatoes

Wine and a brown sugar-butter sauce make this an outstanding vegetable dish to serve with roast beef or steak. If you do not have fresh basil, substitute chopped parsley.

4 large tomatoes
4 tablespoons butter or margarine
2 tablespoons brown sugar
1 teaspoon salt
½ cup white or rosé wine
Pepper
Chopped fresh basil

1. Slice tomatoes into ¼ inch rounds.

2. In large skillet, combine butter and brown sugar and stir until melted. Add salt.

3. Add tomato slices and cook quickly on both sides. Add wine and simmer 1 to 2 minutes or until tomatoes are heated through, basting with the wine sauce.

4. Sprinkle with pepper and chopped basil. Serve in small dishes. Serves 6 to 8.

Baked Macaroni and Tomatoes

Tomatoes, macaroni, and Cheddar cheese combine nicely in this dish. I serve it with broiled hamburgers and a fresh fruit salad. The tomatoes eliminate the need for another vegetable.

1 package (8 ounces) macaroni
3 tablespoons chopped onion
2 tablespoons butter or margarine
1 medium bell pepper, chopped
2½ cups peeled, seeded, and chopped fresh tomatoes
1 teaspoon salt
½ cup grated sharp Cheddar cheese

1. Cook macaroni in boiling salted water until tender. Drain thoroughly.

2. Cook onion and bell pepper in butter until tender. Add tomatoes and simmer about 5 minutes.

3. Add drained macaroni and salt. Mix well and simmer 10 minutes longer.

4. Arrange half of macaroni-tomato mixture in bottom of a greased casserole. Sprinkle with half of the grated cheese. Add remaining macaroni-tomato mixture and sprinkle top with remaining grated cheese.

5. Bake in a 400° oven until cheese is nicely browned, about 15 minutes. Serves 6.

Scalloped Tomatoes

Buttered bread crumbs and Tabasco sauce add greatly to this tasty dish. During the winter, when fresh tomatoes are not available, I substitute canned whole tomatoes, but drain off a bit of the liquid.

1 tablespoon butter or margarine
2 teaspoons minced onion
¼ cup dried green pepper
¼ teaspoon Tabasco sauce
3 cups peeled and quartered tomatoes
1 teaspoon sugar
1 teaspoon salt
1 tablespoon flour
½ cup buttered bread crumbs

1. Melt butter in saucepan. Add onion and green pepper; cook until tender, but not brown.

2. Stir in Tabasco sauce; add tomatoes, sugar, salt, and flour. Bring to a boil; reduce heat and simmer 10 minutes.

3. Turn into a greased 1-quart casserole. Sprinkle with buttered bread crumbs. Bake in 375° oven until crumbs are nicely browned, about 15 minutes. Serves 4.

Vegetable Medley

During the summer when our garden abounds with fresh vegetables, I like to make up combinations of vegetables and seasonings, such as this one. Use this as a basic recipe; add or delete other vegetables, depending on what is in season at the time—bell peppers, carrots. It always tastes different and is a great dish to serve with simply prepared meats.

1 large onion, sliced
1 garlic clove, minced
¼ cup minced parsley
2 teaspoons salt
½ teaspoon each pepper, ground thyme, and dried basil (substitute 1 tablespoon fresh basil leaves, if available)
2 tablespoons cooking oil
1 pound green beans, cut
3 large tomatoes, quartered
2 cups diced summer squash, sliced 1 inch thick

1. Cook onion, garlic, parsley, and seasonings in oil in a large skillet for 3 minutes.

2. Add beans with ½ cup water and simmer, covered, for 10 minutes.

3. Add remaining vegetables, cover and cook until tender but crisp. Don't overcook. Makes 4 to 6 servings.

◁

Vegetable Medley (recipe column 3) varies with the season.

Main dishes and casseroles

For elegant entertaining or simple family-fare, tomatoes will add new taste appeal to many favorite casseroles. Following are some of my favorites.

Tomato Fish Stew

My Italian grandmother always served a thick, robust fish stew (actually more like a soup) during the Lenten Season. I've changed her recipe a bit by using a wider variety of fish and lacing the stew with white wine just before serving. Hot toasted French bread, for dunking, and a tossed green salad, complete the meal.

6 slices bacon, diced
¾ cup chopped onion
1 clove garlic, minced
¾ cup bias cut celery
¼ cup diced green pepper
2 cups peeled and chopped tomatoes
1 cup tomato sauce
1 cup water
3 potatoes, peeled and cut into large chunks
2 teaspoons salt
⅛ teaspoon pepper
½ teaspoon oregano
¼ teaspoon basil
1 bay leaf
1 pound halibut fillets
½ pound fillet of sole
½ pound red snapper
¼ pound codfish
¼ pound shrimp, raw and peeled
1 cup white wine
Parsley for garnish

1. Fry bacon in large saucepan until limp.

2. Add vegetables, except potatoes, and cook until translucent.

3. Add tomatoes, tomato sauce, water, potatoes, and seasonings.

4. Cover and cook about 20 minutes, or until potatoes are tender.

5. Cut fish into bite-size pieces and add to tomato mixture.

6. Cover and continue cooking 5 to 8 minutes, until fish is done. Remove bay leaf.

7. Add wine, sprinkle with chopped parsley garnish. Serves 6.

Tomato Rabbit

A favorite combination of mine is cheese and tomatoes. In this dish I sometimes sprinkle buttered bread crumbs over the tomatoes just before broiling them. Also, adding 2 tablespoons of dry sherry to the rabbit just before serving gives it a nice nutty flavor.

3 tomatoes, cut in half crosswise
Salt, pepper, and sugar to taste
Bread crumbs (optional)
2 tablespoons butter
1 tablespoon prepared mustard
½ teaspoon dry mustard
½ teaspoon paprika
¼ teaspoon salt
¼ teaspoon curry powder
2 teaspoons Worcestershire sauce
1 pound processed American or Cheddar cheese, cut into ½-inch cubes
½ cup beer
2 egg yolks
¼ cup cream
2 tablespoons dry sherry (optional)
English muffins or toast triangles

1. Sprinkle tomatoes with salt, pepper, sugar, and breadcrumbs. Broil until tender and heated through, about 10 minutes. Keep warm while preparing sauce.

2. In the top part of a chafing dish or double boiler place butter to melt. Add prepared mustard, dry mustard, paprika, salt, curry powder, and Worcestershire sauce. Stir well.

3. Add cheese, stirring constantly until melted. Keep water in the lower

Creamy Tomato Rabbit is poured over fresh tomato halves and English Muffins.

Tomato Fish Stew combines fish and a variety of vegetables in a rich tomato sauce.

part of chafing dish or double boiler simmering, but not boiling. If it boils rapidly, it may cause the cheese to become stringy.

4. When cheese has melted, add beer and stir through. Continue cooking until very hot and mixture is thoroughly blended.

5. Beat egg yolks with cream and slowly add to the cheese, stirring constantly.

6. Remove from heat, stir in sherry, if desired.

7. Place broiled tomatoes on top of muffins or toast and pour hot rabbit over. Serve immediately. Serves 6.

Spareribs, Tomatoes, and Green Beans Casserole

This is a very old and delicious recipe that has been in our family for many years. It is a one-pot meal that I prepare often during the summer, when both tomatoes and green beans are in season. Fresh tomatoes mingled with the meat juices transform the liquid into a good, rich gravy.

3 pounds lean, meaty spareribs (have butcher cut rack in half, lengthwise)
1 onion, chopped
1 tablespoon chopped parsley
1 teaspoon oregano
2 cups warm water
1 cup peeled and chopped tomatoes
1 pound green beans
4 potatoes, peeled and quartered
Salt and pepper to taste

1. Cut spareribs into individual pieces; and sauté in a large pot or Dutch oven, together with onion, salt, pepper, parsley, and oregano.

2. When onions are translucent and spareribs are lightly browned, add water and tomatoes and cook for 20 minutes.

3. Add green beans, cover, and cook for another 20 minutes.

4. Add potatoes, cover, and cook an additional 20 minutes or until potatoes are tender. If necessary, remove cover towards end of cooking period and allow gravy to thicken. Salt and pepper to taste. Serves 6.

Fried Green Tomato Slices

I've tried several ways of frying green tomatoes and this method, by far, is the best. The breadcrumbs take on a rich, golden color and become crispy, while the tomatoes inside stay firm, but tender.

6 medium-sized green tomatoes, sliced 1/4 inch thick
1/2 cup all-purpose flour
2 beaten eggs
1 cup seasoned bread crumbs (season with salt, pepper, parsley, chopped onion, and dried oregano)

Green tomato slices are dipped in flour, egg, and seasoned bread crumbs . . .

. . . then fried slowly until crispy and golden brown.

1. Dip tomato slices first in flour, then in egg, then in seasoned bread crumbs.

2. Cook in a small amount of oil, in a large skillet, over medium heat, until lightly browned, about 3 minutes on each side. Tomatoes should be tender when pierced with a fork, but not overdone.

3. Drain on absorbent paper. Serve warm, with lemon wedges, if desired.

Italian Spaghetti Sauce

I always have a container of spaghetti sauce on hand in my freezer. It is what I call my "convenience food." If I'm preparing the sauce for freezing, I omit the meat. This way I can use it in any number of ways—for Swiss steak, as a pizza sauce, for lasagna, in eggplant Parmesan, or by adding clams, chicken, or seafood, as another type of spaghetti sauce.

3 tablespoons olive oil
1 clove garlic, minced
1 celery stalk, chopped
1 carrot, chopped
1 large onion, chopped
1 small bell pepper, chopped
1/4 cup chopped parsley
1/2 pound ground beef
2 cups peeled, chopped tomatoes
1 can (8 ounces) tomato sauce
2 tablespoons tomato paste
1 cup water
1 1/2 teaspoons oregano
1/4 teaspoon thyme
3 fresh basil leaves or 1/2 teaspoon dried
1 bay leaf
1 can (3 ounces) chopped mushrooms, including liquid
1 tablespoon salt
1/2 teaspoon pepper
1 teaspoon sugar
1/2 cup Burgundy wine

1. Sauté vegetables (except tomatoes), parsley, and meat in olive oil until vegetables are tender and meat is lightly browned.

2. Add tomatoes, tomato sauce, tomato paste, and water, and cook together gently for a few minutes.

3. Add herbs, mushrooms, and seasonings and simmer, uncovered, for 2 to 2 1/2 hours, or until thickened. Add wine and adjust seasonings, if necessary; remove bay leaf. Makes approximately 2 quarts.

Baked Tomatoes

4 medium to large ripe tomatoes
1-1 1/2 cups unseasoned croutons
Juice of 1/2 lemon
1/2 teaspoon dried basil
1 tablespoon chopped parsley
1 teaspoon fresh chives
1/4 teaspoon garlic salt
Salt and pepper to taste
4 tablespoons grated Parmesan cheese
2 tablespoons butter

1. Remove a slice from the top of each tomato and scoop out the inside to within 1/2 inch of the shell.

2. Chop pulp into bite-size pieces and combine with remaining ingredients, except cheese, blending only until croutons are moistened.

3. Refill tomato shells with mixture, packing lightly.

4. Sprinkle with grated cheese and dot with butter.

5. Place in lightly oiled baking dish and bake in 350° oven for 20 minutes.

6. Remove from oven and place under broiler for additional 5 minutes or until nicely browned on top. Serves 4.

Breads

The following tomato-flavored breads are excellent toasted and add a nice flavor to sandwiches.

Tomato Bread

This bread is a beautiful orange-red color.

1 package active dry yeast
¼ cup warm water (105-115°F.)
¼ teaspoon sugar
2 cups tomato juice, fresh or canned
2 tablespoons butter
2 teaspoons salt
2 tablespoons sugar
6 to 6½ cups all-purpose flour, unsifted

1. Combine yeast, ¼ cup warm water and ¼ teaspoon sugar in a small bowl. Let set until foamy, about 10 minutes.

2. Heat tomato juice and butter in a saucepan just until warm. Butter does not have to melt.

3. In a large bowl combine salt, 2 tablespoons sugar, and 2 cups of the flour. Blend well.

4. Add tomato juice mixture and yeast mixture to flour mixture and beat 50 strokes with a wire whisk or rotary beater.

5. Add another ½ cup flour to mixture and beat another 50 strokes.

6. With a wooden spoon gradually stir in remaining flour, ½ cupful at a time, to make a stiff dough. Dough is ready when it begins to pull away from the sides of the bowl. Don't add more flour than necessary.

7. Turn dough out on a floured surface and knead, adding more flour only as needed, until dough is smooth and satiny—about 10 minutes.

8. Put dough in a greased bowl, turning once to grease top. Cover bowl with a towel and set in a warm place (about 80°) to rise until doubled. Takes about 1½ hours.

9. Punch down; turn out onto floured board and knead slightly to remove air bubbles.

10. Divide into 2 portions. Shape into smooth ovals—pinch bottom seam, turn ends under, and seal.

11. Put in lightly greased loaf pans, seam side down. Cover and let rise again in a warm place until almost doubled (about 45 minutes).

12. Bake in 350° oven about 45 minutes until nicely browned. Remove from pans and cool on wire racks. Makes 2 loaves.

Tomato Egg Braid

Follow directions for Tomato Bread but use 1¾ cups tomato juice. After adding the ½ cup flour (step 5), add 2 beaten eggs and beat an additional 50 strokes, or until well blended. Again follow the Tomato Bread instructions.

Divide the dough for both loaves into 3 equal parts and roll them into strands about 14 inches long. Braid each trio of strands together, pinching ends to seal.

Let rise on lightly greased baking sheets until almost doubled (about 45 minutes). Brush with slightly beaten egg before baking. Bake at 350° for 30 to 35 minutes until nicely browned.

Tomato Caraway Buffet Loaves

Follow directions for Tomato Bread, adding 1 tablespoon caraway seed to the dry ingredients.

After the first rising, divide the dough into 4 equal parts and roll each into long, thin loaves about 12 to 14 inches long. Let rise on lightly greased baking sheets until almost doubled (35 to 45 minutes).

Brush with slightly beaten egg and sprinkle with more caraway seeds before baking. Bake at 350° until lightly browned, about 30 minutes. Makes 4 loaves.

Tomato Herb Bread

Besides being very tasty, this is a pretty bread to look at, with specks of red and green blended through it. I like to use it for cheese or cold breast-of-chicken sandwiches. It can be baked in a variety of pans—loaf pans, 9-inch pie plates or cake

Baked stuffed tomatoes surround roast chicken for an elegant meal.

An assortment of tomato flavored breads (left to right) Tomato Bread, Tomato Herb Bread, Tomato Caraway Buffet Loaf, Tomato Egg Braid.

pans, or two-quart casserole or soufflé dishes.

1 package dry yeast
¼ cup warm water (105-115°F.)
¼ teaspoon sugar
1½ cups milk
3 tablespoons butter
2 teaspoons salt
2 tablespoons sugar
5½ to 6 cups all-purpose flour, unsifted
2 eggs, beaten
1 tablespoon dried onion flakes
1 large tomato, peeled, seeded and chopped
1 teaspoon dried basil
¼ teaspoon marjoram
¼ teaspoon thyme

1. In a small bowl combine yeast, warm water and ¼ teaspoon sugar. Let set until foamy, about 10 minutes.

2. Heat milk and butter in a saucepan just until warm. Butter does not have to melt.

3. In a large bowl combine salt, 2 tablespoons sugar and 2 cups of the flour. Blend well.

4. Add milk mixture to flour mixture and beat 50 strokes with a wire whisk or rotary beater.

5. Add yeast mixture and another ½ cup flour and beat another 50 strokes.

6. Add beaten eggs and beat mixture until thoroughly blended.

7. Add dried onion flakes, tomato, and herbs. Blend well.

8. With a wooden spoon stir in remaining flour, ½ cup at a time, only until mixture becomes a stiff dough and starts to pull away from the sides of the bowl. Don't add more flour than necessary.

9. Turn dough out on a floured surface and knead, adding more flour only as it is needed, until dough is smooth and satiny—about 10 minutes.

10. Put dough in a greased bowl, turning once to grease top. Cover bowl with a towel and set in a warm place (about 80°F.) to rise until doubled. Takes about 1½ hours.

11. Punch down; turn out onto floured board and knead slightly to remove air bubbles.

12. Divide into 2 portions. Shape into ovals or rounds.

13. Put loaves in lightly greased pans. Cover and let rise again in warm place until almost doubled (about 45 minutes).

14. Bake in 350° oven about 35 to 45 minutes, depending upon the size of the pan. Loaves will look nicely browned and start to pull away from the sides of the pan when they are done.

15. Remove from pans and cool on wire racks. Makes 2 loaves.

Slices of fresh tomato and anchovy fillets top home-baked pizza.

Fresh Tomato Pizza

Pizza in our home is usually served as a first course, or with the antipasto. It is made with a basic white bread dough and shaped on a baking sheet. The topping consists of fresh sliced tomatoes, anchovy fillets, and cheese. However, it can be served as a main course by adding toppings of your own choice—thin-sliced salami, minced ham, shrimp, olives, ground beef, or sausage.

Dough

¼ teaspoon sugar
¼ cup warm water (105-115°F.)
1½ teaspoons dry yeast
1½ cups all-purpose flour
1 teaspoon salt
1 tablespoon butter

1. Dissolve sugar in warm water and sprinkle yeast on top. Set aside for 10 minutes, or until foamy.

2. Sift flour and salt into a bowl. With pastry blender cut in butter until well blended.

3. Add yeast mixture to flour and mix to a dough. Add more flour if dough is sticky.

4. Turn out onto floured surface and knead until smooth and elastic, about 10 minutes.

5. Put into oiled bowl; turn once to coat all sides. Cover and let rise in a warm place (80°F.) until doubled in bulk.

6. Punch down; turn out onto floured board and knead slightly to remove air bubbles.

7. Roll into a circle ¼ inch thick and 12 inches wide. Place on oiled baking sheet.

Topping

Olive oil
Salt and pepper
4 ounces mozzarella cheese, thinly sliced or shredded
4 medium tomatoes, peeled and sliced
12 anchovy fillets
Oregano

1. Brush top of dough base lightly with olive oil. Sprinkle with salt and pepper.

2. Arrange cheese and tomato slices over dough. Lay anchovy fillets between tomato slices. Sprinkle generously with oregano.

3. Bake in 450° oven about 20 minutes until crust is golden brown. Cut into 6 wedges and serve immediately. Serves 6 as an appetizer, or 2 as a main course.

Desserts

The green tomato plays a surprising part in the following three recipes.

Green Tomato Pie

This pie is often compared to rhubarb or tart green apple pie. Actually, it has a taste all its own. I like to serve it slightly warmed, with a scoop of vanilla ice cream. Use your favorite pastry recipe. A flaky, tender pie crust is best with any fruit pie.

Filling

6 cups sliced green tomatoes
Boiling water
1 cup sugar
¼ teaspoon salt
3 tablespoons flour
¼ teaspoon ground nutmeg
¼ teaspoon cinnamon
Small pinch ground cloves
Grated rind and juice of 1 lemon
2 tablespoons butter or margarine
Pastry for 2-crust 9-inch pie

1. Wash tomatoes; do not peel. Slice ⅛ inch thick into bowl. Pour on boiling water to cover; let stand 3 minutes. Drain.

2. In a small bowl combine sugar, salt, flour, and spices. Combine lemon juice and rind in another bowl.

The versatile green tomato goes into an assortment of mouth-watering desserts (clockwise from bottom left): Green Tomato Pie, Mock Date Bars, Mystery Fudge Cake.

3. Fill pastry-lined pie dish with layers of green tomatoes, sprinkling each layer with sugar mixture, dots of butter, and small amounts of lemon mixture.

4. When pan is filled, arrange pastry strips in a lattice pattern over filling. Moisten edges to seal and flute or press together with a fork.

5. Bake in 450° oven for 8 to 10 minutes. Then reduce heat to 375° and bake until tomatoes are tender, about 40 minutes.

6. Remove to a wire rack to cool.

Mystery Fudge Cake

This is about the moistest chocolate fudge cake I have ever tasted. It keeps for days in the refrigerator, without losing its freshness. The mystery—green tomatoes! Sounds strange, I know, but when blended with all the other ingredients, there is absolutely no way one can detect the secret ingredient. It really is delicious and I wouldn't hesitate a minute to serve it to my most discriminating guests.

2½ cups regular, all-purpose flour
½ cup cocoa
2½ teaspoons baking powder
2 teaspoons baking soda
1 teaspoon salt
1 teaspoon cinnamon
¾ cup butter or margarine
2 cups sugar
3 eggs
2 teaspoons vanilla
2 teaspoons grated orange peel
2 cups coarsely grated green tomato
1 cup finely chopped walnuts
½ cup milk

1. Combine dry ingredients in a large bowl and set aside.

2. In another bowl beat together butter and sugar until smooth and blended.

3. Add eggs one at a time and beat well after each addition.

4. With a wooden spoon, stir in vanilla, orange peel, and green tomatoes.

5. Stir the nuts into the sifted dry ingredients and add to tomato mixture alternately with milk.

6. Pour batter into a greased and floured 10-inch tube or bundt pan.

7. Bake at 350° until a wooden pick inserted in center comes out clean, about 1 hour.

8. Cool in pan 15 minutes then turn out onto a wire rack and cool thoroughly.

9. Sprinkle with powdered sugar before serving.

Mock Date Bars

This is a delightfully chewy bar-type cookie. It's also very economical since it doesn't use dates at all. The filling is made with ground green tomatoes and walnuts, but I doubt if anyone would ever know the difference. The whole-wheat flour in the crust makes it extra wholesome and gives it a nutlike flavor.

Filling

2 cups ground green tomatoes
½ cup +2 tablespoons sugar
Juice and grated peel of 1 orange
1 tablespoon lemon juice
1 teaspoon cinnamon
¼ teaspoon cloves
½ teaspoon salt
½ cup chopped walnuts

1. Combine all above ingredients, except walnuts, in a saucepan and cook until most of liquid has evaporated and mixture is quite thick. Stir occasionally.

2. Stir in walnuts and cool while preparing crust.

Crust

½ cup butter or margarine
1 cup firmly packed brown sugar
2 teaspoons vanilla
1½ cups sifted whole-wheat flour
1 teaspoon baking powder
½ teaspoon baking soda
1 teaspoon salt
1 cup quick-cooking oatmeal

1. Cream butter, sugar, and vanilla together.

2. Sift together whole-wheat flour, baking powder, baking soda, and salt. Stir into creamed mixture until smooth.

3. Add oatmeal and blend thoroughly.

4. Grease and flour an 8 x 12-inch baking pan. Press half of the mixture firmly into pan. Cover with filling.

5. Place remaining oatmeal mixture over filling. Press down.

6. Bake at 350° for 30 minutes. Cool. Cut into 2-inch squares. Makes 24.

Canning tomatoes

I welcome what other people might consider an overabundance of tomatoes. A large crop means enough canned tomatoes to see me through the rest of the year. I'll match my canned tomatoes with any you can buy in the store.

I use two methods of canning tomatoes: the cold-pack method, and the hot-pack. The difference is this: cold-pack tomatoes are peeled, dropped into the jars whole, and processed. They retain their shape after canning. Hot-pack tomatoes are

For a good seal

1. Check the jar's edge for cracks or chips. If you find any, discard the jar.

2. Fill one jar at a time. Leave ½" of headspace between the tomatoes and the jar rim.

3. Wipe the rim with a clean, damp cloth.

4. Seal the jar immediately with a lid and . . .

5. Put on a metal ring and screw it down tightly with your hand *only*.

6. In about 12 hours, check the seal by pressing the center of the lid. It should be drawn down tightly . . .

If it bounces up and down to the pressure of your finger the jar is *not* properly sealed.

packed into the jars hot; that is, they are peeled, quartered, heated to the boiling point, and then packed into the jars. The result is more like a sauce.

Equipment needed

✔Canning jars (I find the wide-mouth quart size the most practical.)

✔Metal rings and lids to fit the jars. (New lids must be used for each canning. Rings may be reused as long as they are in good condition and free from rust.)

✔Water-bath canner with a metal rack inside to hold the jars, and a tight-fitting cover, or any large kettle that is deep enough to permit water to cover the jars at least one inch over the tops, with a little extra space to allow for boiling. You can improvise a rack with strips of wood or wire, or use any size rack that will allow the boiling water to circulate under the jars.

✔Jar lifter, designed to wrap around jars while filling, and to grasp neck of jars when lifting in and out of the boiling water.

✔Miscellaneous utensils: Canning funnel, paring knife, long knife or narrow spatula, clean cloth, plastic or wooden spoon, clock or kitchen timer, measuring spoons, and towels or cooling rack.

The ingredients

The ingredients are simple: tomatoes and a little salt. Select the tomatoes carefully. Use only fresh, firm, ripe tomatoes, free from any decayed spots. After the jars are filled with tomatoes, add salt at the rate of ½ teaspoon per pint, or 1 teaspoon per quart, depending on the size jars you use.

Here's how

1. Set out all equipment and utensils within easy reach. Fill kettle ⅔ with water and put on high heat to boil. Keep a tea kettle filled with hot water, which will be needed later when all the jars are filled and sealed.

2. Examine tops of jars to make sure there are no nicks, cracks, or sharp edges that would prevent a perfect seal. Imperfect jars should not be used for canning.

3. Wash the jars thoroughly in hot, soapy water. Rinse completely. Keep jars in hot water until you are ready to use them.

4. Select enough tomatoes to make one canner load—most canners hold seven jars. It takes approximately 2½ to 3 pounds of tomatoes for each quart jar.

5. Peel tomatoes according to directions on page 72. Cut out core, remove or trim away any green spots.

To cold-pack tomatoes

1. Remove one jar from water and drain. Place funnel on jar. Drop whole tomatoes into jar, Cut very large tomatoes into quarters, if necessary.

2. Remove funnel and press tomatoes gently to fill space, to within ½ inch of the top of the jars. To get the proper headspace, add a little more tomato, or pour off a little juice. Do not add water.

3. Add 1 teaspoon salt to each quart jar, or ½ teaspoon to each pint.

4. Run a long knife or narrow spatula between tomatoes and jar to release any trapped air bubbles.

5. Wipe the top of the jar with a clean, damp cloth and check that no seeds or bits of tomato are clinging to the rim or outer threads.

6. Remove one lid from the boiling water and place it flat on top of the jar, with the sealing compound next to the glass. Place metal ring over lid and screw it down *firmly.*

7. When all jars are filled, stand them on the rack in the canner. Water should cover jars by at least 1 inch. Add more water (from the tea kettle) if necessary.

8. Put cover on canner. Bring water to a rolling boil. Begin counting processing time—45 minutes for quarts; 35 minutes for pints. Reduce water to hard simmer and hold it there the entire time.

9. Using the jar lifter, remove jars from the canner and set on a rack or

A jarlifter is a necessary tool when handling hot canning jars.

folded cloth, allowing space between jars for air circulation. Do not set jars in a draft or cover them. Do not tighten metal rings.

10. When completely cool (about 12 hours), test for seal by pressing down on the center of the lid. If the dome is down and gives out a clear, ringing sound when tapped with a spoon, the jar is sealed. Remove bands and store jars in a cool, dark, dry place.

To hot-pack tomatoes

Peel and quarter tomatoes. Put in large pan and heat slowly to the boiling point. Stir frequently to avoid scorching. Fill jars while tomatoes are still hot.

Follow the procedure for cold-pack tomatoes with the following change:

Subtract 10 minutes from the processing time, for both quarts and pints.

Jams, butter, marmalade, and preserves

In spring, when planning our vegetable garden, we have no worry about overplanting tomatoes. I always make sure we plant enough so that I'll have a good crop of both red and green tomatoes to put up the jams, marmalade, and other good things my family enjoys so much.

Red Tomato Jam

This is a beautifully clear, red jam that uses many of the fully or overripe tomatoes. I always make this toward the end of the season when tomatoes are ripening quicker than I can use them fresh.

3 pounds (6 large) fully ripe tomatoes
1 lemon, thinly sliced
1 box (2½ ounces) powdered fruit pectin
4 cups sugar

1. Scald, peel, and quarter tomatoes, removing stems and cores. Remove seeds and drain off juice, reserving only the pulp.

2. Bring to boil in a large kettle. Reduce heat and simmer, uncovered, for 8 to 10 minutes. Measure; there should be 3 cups.

3. Put tomatoes, lemon, and pectin in a large kettle. Bring to a full boil, stirring constantly. Add sugar and boil rapidly for another 2 minutes.

4. Cool for 5 minutes, stirring occasionally.

5. Fill sterilized jars and seal. Process in boiling water bath for 10 minutes. (See pages 90-92.) Makes 2 pints.

Green Tomato Jam

This is a jam I make early in the season. Pick the tomatoes when they

Tomatoes, both red and green, are used in making jams, marmalade, preserves, and butter.

are of a good size and about one week before turning red.

6 cups (about 2½ pounds) chopped green tomatoes
Hot water
4 cups sugar
2 lemons, thinly sliced

1. Cover tomatoes with hot water and boil for 5 minutes.

2. Drain and add sugar. Let stand for 3 hours or longer.

3. Drain syrup into kettle; bring to a boil and cook rapidly until quite thick.

4. Add tomatoes and sliced lemon and cook until thick and clear, about 10 minutes.

5. Pack into sterilized jars and seal. (See pages 90-92.) Makes 2 pints.

Green Tomato-Citrus Marmalade

This is our favorite because it is not quite as sweet as most marmalades. The green tomatoes soften considerably after cooking and the jam becomes very thick. Remember to stir constantly as it thickens rapidly once it comes to a full boil.

1½ quarts green tomatoes, thinly sliced
3 cups sugar
½ teaspoon salt
2 lemons
2 oranges

1. Mix tomatoes, sugar, and salt in a large bowl.

2. Peel lemons and oranges. Boil peel for 8 minutes in enough water to cover. Drain off and discard the water. Cut peel into thin strips.

3. Slice lemon and orange pulp very thin and remove seeds.

4. Combine peel, citrus pulp, and tomato mixture in a large kettle. Heat to boiling and cook rapidly, stirring constantly until thickened, about 45 minutes.

5. Pour into sterilized jars and seal. Process in boiling water bath for 10 minutes. (See pages 90-92.) Makes 2 pints.

Tomato-Apple Butter

Tomatoes and apples combine to make a very tasty butter. It's delicious on homemade biscuits and very good in peanut butter sandwiches.

3 pounds (about 6) ripe tomatoes
3 green apples
2 tablespoons lemon juice
1 teaspoon salt
1½ teaspoons cinnamon
½ teaspoon ground cloves
¼ teaspoon allspice
¼ teaspoon nutmeg
2½ cups brown sugar

1. Scald, peel, and quarter tomatoes. Cook, covered, until mushy, stirring occasionally. Press through a sieve to remove any seeds. Measure; there should be about 4 cups.

2. Peel, core, and quarter apples. Cook in ½ cup water until mushy, about 15 minutes.

3. Combine tomato purée and apple pulp in a large kettle. Add remaining ingredients. Bring to a boil and simmer, uncovered, stirring frequently for 45 minutes, or until thick.

4. Fill sterilized jars and seal. Process in boiling water bath for 10 minutes. (See pages 90-92.) Makes 2 pints.

Green Tomato Mincemeat

I can hardly wait for my green tomatoes to get large enough so that I can make this marvelous mincemeat. Although it requires 3 hours of simmering and must be stirred often, it is not as difficult to make as the time involved would suggest. It is well worth the time spent. I use it in cookies, cakes, and pies, or warmed and served over vanilla ice cream. A bit of rum or brandy added to the warmed mincemeat gives it a nice festive touch.

6 cups chopped, peeled apples
6 cups chopped green tomatoes
4 cups light brown sugar
1¼ cups cider vinegar
2 cups golden raisins
3 teaspoons cinnamon
1 teaspoon ground cloves
¾ teaspoon allspice
¾ teaspoon mace
2 teaspoons salt
½ teaspoon freshly ground black pepper
½ cup butter

1. Mix the apples and tomatoes together. Add the remaining ingredients except the butter.

2. Bring gradually to a boil, in a large pot, and simmer for three hours, stirring often.

3. Add the butter and mix well. Spoon into hot, sterilized, canning jars and seal.

4. Process in boiling water bath for 25 minutes. (See pages 90-92.)

5. Remove from kettle and allow to cool.

6. Store in a cool, dark, dry place. Makes about 5 pints.

Piccalilli

1 quart cabbage, chopped
1 quart chopped green tomatoes
1 cup chopped celery
2 large onions, chopped
2 sweet red peppers, seeded and chopped
2 green peppers, seeded and chopped
¼ cup salt
1 stick (2 inches) cinnamon
1 teaspoon each whole cloves and whole allspice
1½ cups cider vinegar
1½ cups water
2 cups brown sugar, firmly packed
1 teaspoon dry mustard
1 teaspoon turmeric
Dash Tabasco sauce

1. Combine vegetables and salt; cover and let stand overnight.

2. Drain off as much liquid as possible, pressing through a clean, thin white cloth, if necessary.

3. Place cinnamon, cloves, and allspice in a cheesecloth bag.

4. Place vegetables, spice bag, vinegar, water, sugar, dry mustard, turmeric, and Tabasco sauce in a large kettle.

5. Bring to a boiling point; reduce heat and simmer for 20 minutes.

6. Pour into sterilized jars and seal. Process in boiling water bath 5 minutes. (See pages 90-92.) Makes 6 pints.

Green Tomato Relish

2 cups (4 or 5 medium) ground green tomatoes
2 cups (2 medium) peeled and ground cucumbers
2 cups (4 medium) ground onions
3 medium-size tart apples, peeled and ground
1 green pepper, seeded and ground
2 small, sweet red peppers, seeded and ground
4 cups water
1½ tablespoons salt
2 cups white sugar
2 cups cider vinegar
1 tablespoon mustard seed
6 tablespoons all-purpose flour
1 tablespoon dry mustard
¼ teaspoon turmeric

1. Mix tomatoes, cucumbers, onions, apples, peppers, water, and salt in a large kettle. Let stand for 24 hours; drain off liquid.

2. Add sugar, 1½ cups of the vinegar, and mustard seed. Bring to a slow boil and cook for 30 minutes, stirring occasionally.

3. Make a paste of flour, dry mustard, turmeric, and remaining ½ cup vinegar. Stir into boiling mixture.

4. Simmer for additional 30 minutes or until mixture has thickened, stirring occasionally.

5. Fill sterilized jars and seal. Process in boiling water bath for 10 minutes. (See pages 90-92.) Makes 4 pints.

Italian Spaghetti Sauce (recipe page 86) uses an assortment of herbs and seasonings.

Perfect go-togethers—juicy slices of fresh tomato, sweet red onions, and leaves of fresh basil.

Companion herbs

The following herbs are the ones I reach for most often when there are tomatoes in the kitchen. Six herbs are just a starting point. You'll find the value of others by experimenting, as I did. That's half the fun of cooking.

Basil. The flavor of common sweet basil is like spicy cloves. It is one of the most delicate herbs and may be used more generously than the lustier ones. Whether fresh or dried, basil is the perfect compliment to any tomato dish; the two flavors seem to have been made for each other.

There are two basic kinds of basil—sweet *(Ocimum basilicum)* and bush *(Ocimum minimum).*

Sweet basil is an annual, growing one to two feet tall, with large two-inch, glossy, dark green leaves. This is the basil most commonly grown for cooking purposes.

Bush basil is a compact form that grows six to 12 inches tall, with tiny whitish flowers. The leaves are smaller than those of sweet basil, and the flavor milder. The most famous use for basil is in pesto sauce.

Pesto Sauce

One of the great sauces of Italy, this easily prepared combination of fresh basil, garlic, olive oil, and Parmesan cheese can be prepared in the blender and frozen in small individual mounds, to be used in a wide variety of ways.

2 cups firmly packed basil leaves, rinsed well
3 cloves garlic, peeled
½ cup olive oil
1 cup shredded Parmesan cheese

Put basil in a blender jar; add garlic and olive oil and blend at high speed until a very coarse purée is formed. Add cheese, a small amount at a time, until thoroughly blended. Makes about 1½ cups. Will keep for one week in a covered container in the refrigerator. For longer storage, drop into small mounds on foil, freeze, then package in a plastic bag, tightly wrapped.

To use, blend small amounts with butter, mayonnaise, or Italian-type salad dressing for use on fresh or cooked vegetables. Mix with cooked noodles, soups, stews, or cold seafood salads.

Coriander is a small plant in the parsley family, with a flavor resembling a combination of lemon peel and sage. It is often sold as Chinese parsley or cilantro. The fresh leaves are a tasty and fragrant addition to mixed green salads and sliced fresh tomatoes.

Marjoram. This tender perennial is a member of the mint family. It has a fragrant aroma and a spicy taste similar to that of sage, though considerably less strong.

Dried marjoram is excellent used in a marinade for fresh sliced tomatoes.

Marinated Tomato Slices

6 medium tomatoes, peeled and sliced
¾ cup salad oil
¼ cup wine vinegar
2 tablespoons chopped green onion
1 teaspoon salt
Dash ground black pepper
2 teaspoons crushed dried marjoram

Place tomato slices in a deep bowl. Combine remaining ingredients, shaking well to blend thoroughly. Pour over tomatoes, cover, and chill for several hours, turning occasionally. Serves 6.

Oregano is among the more potent herbs and can be used in all tomato dishes. Fresh or dried leaves and tops are used. They have a sweet, aromatic flavor like that of sweet marjoram or thyme, though oregano is stronger and should be used with care. It is a favorite with Mexican and Italian cooks, and is often referred to as the "pizza herb."

Parsley is our most common herb. There are three main types of parsley grown in this country—curled, plain-leaved, and turnip-rooted. Very popular is the moss curled variety, which has very dark green leaves, deeply curled, and is splendid for seasoning and general culinary decoration. Parsley leaves have a familiar, refreshing taste and aroma.

Bouquet garni

Parsley is an essential ingredient in a bouquet garni—a number of herbs and spices tied together in a cheesecloth bag so that they may be easily discarded after cooking. A bouquet garni are generally used in slow-cooking dishes such as soups and stews.

A basic bouquet garni is made with a few parsley sprigs, celery leaves, onion slices, and a sprig of thyme. However, depending on the dish to be flavored, any one or combination of other herbs and spices may be added—basil, garlic, chives, rosemary, tarragon, whole cloves.

Thyme, a member of the mint family, is a moderately potent herb. It has a pungent flavor and sweet fragrance. It is one of the traditional herbs of Creole cuisine. Also, it is an essential ingredient in a bouquet garni.

Tomato notes — last words

The number of days in the year that you can pick ripe fruit from vines in the garden is limited. Yet, the use of ripe tomatoes is a year-round affair.

Almost all store-bought fresh tomatoes are ripened *off* the vine.

The problems of transportation, storage, and ripening have been constantly improved, but the fully ripe tomato cannot be shipped. As stated in the United Fresh Fruit & Vegetable Association's bulletin on tomatoes: "The ideal tomato, from the consumer's viewpoint, is one that is full size, vine-ripened, unblemished, and characteristically at the red-ripe stage or anything near that stage. Such fruit is too tender even to stand commercial harvesting, let alone packing and shipping. (Note: So-called 'vine-ripe' tomatoes are not fully vine-ripe but are harvested at the turning or pink stage.) On the other hand, if a grower harvests tomatoes at the stage where they are most resistant to shipping injury, they lack juiciness and flavor."

A "mature-green" tomato has a glossy appearance, but no red color. The gel in the seed cavity is well formed and the seeds slip off the edge of a sharp knife when the tomato is sliced.

Commercially grown tomatoes are considered vine-ripe when color begins to show at the blossom end. Whether mature-green or vine-ripe, tomatoes should be allowed to ripen fully for best flavor.

Ripening

The best temperature range for ripening a tomato is 60° to 70° F. Flavor will be impaired when tomatoes are exposed to temperatures below 55° F. High temperatures (above 80° F.) also prevent good color and flavor, and increase the chance of decay.

Moderate light will speed ripening; too much sunlight prevents development of normal, even color.

Don't expect *immature* green tomatoes to ripen; they have a bitter taste and are likely to rot if kept for any length of time. When selecting green tomatoes for cooking, select those of mature size, that would, under ordinary conditions, be about a week from turning red. (See description of mature-green tomatoes above.)

Storage

For the best flavor, ripe tomatoes should not be stored in the refrigerator for any length of time. As stated above, temperatures below 55° F. are damaging to the fruit. (Average refrigerator temperature is 45° F.) You will get top flavor if tomatoes are stored in a cool place as near to 60° F. as possible.

It is best to make a practice of peeling and cutting tomatoes just before using. If it is necessary to store peeled tomatoes, wrap in wax paper, plastic wrap, or foil, and refrigerate.

Use the following guide when storing tomatoes.

✔ Fresh, ripe, raw: cool place 60° F., or refrigerate no more than 2 to 3 days.

✔ Fresh, cooked; canned, opened: refrigerator shelf 4 to 5 days.

✔ Canned kitchen shelf, 1 year.

✔ In cooked dishes, stored in refrigerator frozen-food compartment, prepared for freezing: 2 to 3 months.

✔ In cooked dishes stored in freezer at zero degrees: 1 year.

As one good gardener writes us, the best storage at the end of the growing season is as follows: "We eat fresh tomatoes into December in spite of early October frosts. Just before the first frost, we tear the whole vine out of the ground and hang it (actually 8 to 10 vines) upside down in the coldest part of our basement. This is the best method of post-frost storage I have run into. The fruit slowly ripens, and you pick them as they ripen."

Nutritional value

Tomatoes are a good source of vitamins and minerals, especially vitamins A and C. A 3½-ounce serving of raw tomato provides 1500 International Units of vitamin A, and 25 milligrams of vitamin C. Small amounts of other vitamins and minerals are present too. Unlike green vegetables, tomatoes lose very little of their nutrients in cooking.

If you want to be slim and trim, add tomatoes to your diet. A 3½-ounce serving of raw tomatoes yields 22 calories; a one-cup serving of canned tomato juice yields 45 calories.

How much?

When measuring tomatoes, either fresh or canned, a handy guide is as follows.

✔ One pound of tomatoes consists of 2 large, 3 medium (2 x 2½ inches) or 4 small tomatoes.

✔ 8-ounce can equals 1 cup cooked tomatoes.

✔ 12-ounce can equals 1½ cups cooked tomatoes.

✔ 16-ounce can equals 2 cups cooked tomatoes.

✔ 28-ounce can equals 3½ cups cooked tomatoes.

✔ 46-ounce can of juice equals 5¾ cups.

✔ One bushel of tomatoes (50 pounds) will make 20 quarts of cooked tomatoes.

Cooking tomatoes

To insure the best color, flavor, texture, and food value, cook tomatoes only until they are tender. Use a small amount of water; they create their own liquid as they cook. Also, the less water used, the more nutrients will be retained in the finished product.

Use one pound of tomatoes for four half-cup servings of stewed tomatoes. The cooking time for this amount of cut-up tomatoes will be about 10 minutes.

How much acid to add

For insurance that the tomatoes you can are in the low pH range, add ¼ teaspoon citric acid, or 1 tablespoon lemon juice, per pint of tomato product.

The chart below shows what happens to the pH ratings of various tomato varieties when different acids are added. Note that the pH is not significantly lowered by the addition of 1 teaspoon lemon juice per pint.

Tomato Variety	Original pH	pH after adding ¼ tsp. citric acid per pint	pH after adding 1 tsp. lemon juice per pint	pH after adding 1 tbsp. lemon juice per pint
Ace	4.4	4.1	4.3	4.2
Ace 55 VF	4.6	4.2	4.5	4.3
Big Early	4.4	4.0	4.3	4.1
Big Girl	4.5	4.1	4.4	4.2
Garden State	4.5	4.2	4.4	4.3
Valiant	4.4	3.9	4.3	4.1

List of catalog sources

In the chapter on varieties, pages 38 to 57, we indicate the catalog sources on all varieties that may be hard to find. They are listed in this fashion:

Redpak (13).

These are the same as the numbers in parentheses below—showing you which seed companies to write to for the catalog you want.

In our listing of available catalogs, we have given a short description of each, showing the number of pages total, the number of pages devoted to vegetables. By reading these descriptions, you can determine the companies that are specialists and those that cover the entire spectrum of garden materials.

Many of these seed catalogs are the equivalents of garden books containing good reference material. They reward the careful reader.

Look especially for the new varieties of bush-type cucumbers, squash and pumpkins. Also the midget varieties of carrots, eggplant, cucumbers, watermelons, cantaloupes, and lettuce.

Our catalog search was made with 1976 information. You can expect changes in the catalogs of later dates. What is scarce today may be widely available in future years, but our search will uncover the most likely sources of hard-to-find varieties.

Note that there are many missing numbers in the following list of seed companies. Don't worry about this absence. The missing numbers are assigned to catalogs appearing in other regional editions.

(2) **Rocky Mountain Seed Co.**
1321-27 15th St., Denver, CO 80217
64-page general catalog; vegetables, 20 pages. Features herbs, ornamental grasses, wild flowers.

(3) **Meyer Seed Co.**
600 So. Carolina St., Baltimore, MD 21231
Vegetables, 23 pages.
Features All-America Selections.

(4) **Burgess Seed & Plant Co.**
P.O. Box 82, Galesburg, MI 49053
44 pages, 8½ x 11. Vegetables, 26 pages. Special attention to varieties for Northern States. Many unusual items.

(5) **W. Atlee Burpee Co.**
(Free from your nearest Burpee branch): Warminster, PA 18974; Clinton, IA 52732; Riverside, CA 92502
170 pages, 6 x 9. When seed catalogs are mentioned most people think "Burpee."

(6) **D. V. Burrell Seed Growers Co.**
Box 150, Rocky Ford, CO 81067
96 pages, 8½ x 4¼. Seed growers. Special emphasis on melons, peppers, tomatoes, and varieties for California and the Southwest.

(8) **Jackson & Perkins**
Medford, OR 97501
The well-known name in roses adds the Jackson & Perkins Seedbook to their catalog list. 12 big pages on vegetables in the 40-page catalog.

(9) **Farmer Seed & Nursery Co.**
Faribault, MN 55021
84 pages, 8 x 10. Complete. Special attention to midget vegetables and early-maturing varieties for northern tier of states. Established 1888.

(10) **Henry Field Seed and Nursery Co.**
407 Sycamore St., Shenandoah, IA 51601
116 pages, 8½ x 11. A complete catalog. Wide variety selection. Many hard-to-find items, good tips for vegetable gardeners.

(11) **DeGiorgi Co., Inc.**
Council Bluffs, IA 51501
112 pages, 8½ x 11. Attention to the unusual. Established 1905. 35¢

(12) **Gurney Seed & Nursery Co.**
1448 Page St., Yankton, SD 57078
76 pages, 15 x 20. Emphasis on short-season North-country varieties. Accent on the unusual items.

(13) **Joseph Harris Co.**
Moreton Farm, Rochester, NY 14624
92 pages, 8½ x 11. Vegetables, 39 pages. The look of authority, and so considered, especially in the Northeast.

(14) **Charles C. Hart Seed Co.**
Box 169, Wethersfield, CT 06109
Vegetables, herbs, and flowers, 24 pages, including All-America Selections.

(15) **H. G. Hastings Co.**
Box 4274, Atlanta, GA 30302
64 pages 8½ 11. "85 years of service to the South." Complete southern garden guide.

(16) **J. W. Jung Seed Co.**
Station 8, Randolph, WI 53956
60 pages, 9 x 12. Everything for the garden. Vegetables, 18 pages. Attention to Experiment Station introductions.

(18) **Earl May Seed & Nursery Co.**
Shenandoah, IA 51603
Complete catalog. 80 pages, 9½ x 12½. Wide choice of varieties. Features All-America Selections.

(19) **Nichols Garden Nursery**
1190 No. Pacific Highway, Albany, OR 97321
88 pages, 8½ x 11. Written by an enthusiastic gardener and cook who has searched the world for the unusual and rare in vegetables and herbs.

(20) **L. L. Olds Seed Co.**
2901 Packers Ave., Box 1069, Madison, WI 53701
80 pages, 8 x 10. Carefully written guide to varieties. All-America Selections; vegetables, 30 pages.

(21) **Geo. W. Park Seed Co., Inc.**
Greenwood, SC 29646
122 pages, 8¼ x 11¼. A guide to quality and variety in flowers and vegetables. Includes indoor gardening. The most frequently "borrowed" seed catalog.

(22) **Reuter Seed Co., Inc.**
New Orleans, LA 70119
32 pages, 8 x 10. Vegetables, 16 pages. "Serving the South since 1881."

(23) **Seedway**
Hall, NY 14463
36 pages, 8½ x 11. Vegetables, 19 pages. Informative, straightforward presentation.

(25) **J. L. Hudson**
P.O. Box 1058, Redwood City, CA 94604
112 pages, 5½ x 9. Vegetables, 16 pages. Ask for vegetable catalog—it's free. General catalog, 50¢. Accent on the unusual. Wide selection of herbs.

(26) **R. H. Shumway, Seedsman**
628 Cedar St., Rockford, IL 61101
88 pages, 10 x 13. Complete. Founded 1870. Catalog has maintained some of the "good farming" 1870 look.

(27) **Stokes Seeds**
Box 548 Main Post Office, Buffalo, NY 14240
Also: St. Catherine's, Ontario, Canada
158 pages, 500 different vegetable and 800 different flower varieties. Emphasis on short-season strains. Canadian & European introductions.

(28) **Otis S. Twilley Seed Co.**
Salisbury, MD 21801
64 pages, 8½ x 11. Clear, helpful presentation with special attention to Experiment Station releases, and disease-resistant varieties, for varying climatic conditions.

(30) **Nicholson-Hardie Vegetable and Flower Seeds**
5717 West Lovers Lane, Dallas, TX 75209.
15-page catalog stressing Southern favorites.

(31) **Glecklers Seedmen**
Metamora, OH 43540
4 pages, 8½ x 14. Listings, brief descriptions of unusual, strange vegetables.
Yearly supplements.

(32) **Geo. Tait & Sons, Inc.**
900 Tidewater Dr., Norfolk, VA 23504
Vegetables, 24 of 58 pages. Special varieties and planting information for eastern Virginia and North Carolina.

(37) **Porter & Son, Seedsmen**
Stephenville, TX 76401
30 pages, 12 pages on vegetables with emphasis on Texas varieties.

(44) **Johnny's Selected Seeds**
Albion, ME 04910
28 pages, 5½ x 8½. Heirloom beans and corn; many hard-to-find seeds of Oriental vegetables. 25¢.

(45) **Kitazawa Seed Co.**
356 W. Taylor St., San Jose, CA 95110
One-sheet listing of Oriental vegetables including bitter melon, Japanese pickling melon, gobo, and many varieties of more common Oriental vegetables.

(46) **J. A. Demonchaux Co.**
225 Jackson, Topeka, KS 66603
Gourmet garden seeds from France.
Four-page list of vegetables and herbs.

(47) **Grace's Gardens**
100 Autumn Lane, Hackettstown, NJ 07840
16-page catalog of unusual seeds, "grow a complete 'believe it or not' food garden in less than ¼ acre." 25¢.

(48) **Tsang & Ma International**
1556 Laurel St., San Carlos, CA 94070
One-page leaflet, 16 types of Chinese vegetables, including bitter melon and Chinese okra. $2 minimum order.

(49) **Burnett Brothers, Inc.**
92 Chambers St., New York, NY 10007
48-page general catalog. Flower and vegetable seeds; garden supplies.

(51) **Archias Seed Store Corp.**
Box 109, Sedalia, MO 65301
42-page catalog includes vegetables, flowers, garden aids, roses, fruits, and berries.
13 pages of vegetable seeds.

(52) **Horticultural Enterprises**
Box 34082, Dallas, TX 75234
22-page catalog, 5½ x 8½. Entirely devoted to varieties of peppers and tomatillo.

Throughout the description of varieties we give the names of the originators of the variety. In case you cannot locate the variety in the normal trade channels—seed racks, nursery transplants, or seed catalogs—we suggest that you write to the wholesale firm that is listed as the originator. They will help you find a source.

*(53) **Geo. J. Ball, Inc.**
Box 355, West Chicago, IL 60185

*(54) **Goldsmith Seeds, Inc.**
Gilroy, CA 95020

*(55) **Ferry-Morse Seed Co.**
Box 100, Mountain View, CA 94040
Ferry-Morse varieties are available in many garden stores. They offer a 48-page booklet, *Home Garden Guide*. The cost is 98¢.

*(56) **Petoseed Co., Inc.**
Box 4206, Saticoy, CA 93003

*(57) **T. Sakata Seed Co.**
120 Montgomery St., San Francisco, CA 94101